Universitext

Universitext

Universitext is a series of textbooks that presents material from a wide variety of mathematical disciplines at master's level and beyond. The books, often well class-tested by their author, may have an informal, personal even experimental approach to their subject matter. Some of the most successful and established books in the series have evolved through several editions, always following the evolution of teaching curricula, to very polished texts.

Thus as research topics trickle down into graduate-level teaching, first textbooks written for new, cutting-edge courses may make their way into *Universitext*.

More information about this series at http://www.springer.com/series/223

Robert G. Underwood

Fundamentals of Hopf
Algebras

Robert G. Underwood
Department of Mathematics and Computer Science
Auburn University at Montgomery
Montgomery, AL, USA

ISSN 0172-5939   ISSN 2191-6675   (electronic)
Universitext
ISBN 978-3-319-18990-1   ISBN 978-3-319-18991-8   (eBook)
DOI 10.1007/978-3-319-18991-8

Library of Congress Control Number: 2015939602

Mathematics Subject Classification (2010): 11Axx, 12E20, 13Axx, 16T05, 16T10, 16T15

Springer Cham Heidelberg New York Dordrecht London
© Springer International Publishing Switzerland 2015

Printed on acid-free paper

Springer International Publishing AG Switzerland is part of Springer Science+Business Media (www.
springer.com)

to my mother and father

Preface

The purpose of this text is to provide an introduction to the fundamentals of coalgebras, bialgebras, and Hopf algebras and their applications. These topics are at the forefront of modern algebra today. The target audience for this book is graduate students in mathematics who would like to know more about this fascinating subject. Researchers in bialgebras and Hopf algebras may also find this book useful.

The prerequisites for the book include the standard material on groups, rings, modules, algebraic extension fields, finite fields, and linearly recursive sequences found in undergraduate courses on these subjects. The book may be used as the main text or as a supplementary text for a graduate algebra course. The reader is referred to [Ch79, La84, Ro02], and [LN97] for a review of this material. That said, it has been the intention to make this book as self-contained as possible. Most of the proofs are given with ample details; the desire was to make them as transparent as possible. A few proofs have been omitted since they are beyond the scope of the book, and for these I have provided references.

The book consists of four chapters. In Chap. 1, we introduce algebras and coalgebras over a field K and show that if C is a coalgebra, then its linear dual C^* is an algebra. On the other hand, if A is an algebra, then A^* may not be a coalgebra. If we replace the linear dual with the finite dual $A^\circ \subseteq A^*$ however, then the algebra structure of A yields a coalgebra structure on A°. In the case $A = K[x]$, we show that the collection of linearly recursive sequences of all orders over K can be identified with the coalgebra $K[x]^\circ$. This suggests the novel problem of finding the image of a linearly recursive sequence $\{s_n\}$ under the comultiplication map of $K[x]^\circ$.

In Chap. 2, we treat bialgebras—vector spaces that are both algebras and coalgebras. We show that if B is a bialgebra, then B° is a bialgebra. For $B = K[x]$, we show that there are exactly two bialgebra structures on $K[x]$. It follows that there are exactly two bialgebra structures on $K[x]^\circ$, and so, we can multiply linearly recursive sequences in $K[x]^\circ$ in two ways: one way is the Hadamard product and the other is the Hurwitz product.

We close Chap. 2 with an application of bialgebras to finite automata, formal languages, and the classical Myhill–Nerode theorem of computer science. Specifically, the Myhill–Nerode theorem is generalized to an algebraic setting in which a certain function plays the role of the language and a "Myhill–Nerode bialgebra" plays the role of the finite automaton that accepts the language. Included are several examples of Myhill–Nerode bialgebras. We introduce regular sequences as generalizations of kth-order linearly recursive sequences over the Galois field $\mathrm{GF}(p^m)$.

Chapter 3 concerns Hopf algebras, which are bialgebras with an additional map (the coinverse map) satisfying the coinverse property. We give some examples of Hopf algebras and discuss some of their properties. In many ways, the group ring KG, G a finite group, is the prototypical example of a Hopf algebra in that the group ring is cocommutative and its coinverse has order 2.

We introduce the space of integrals of a Hopf algebra H and Hopf modules over H, and state the Fundamental Theorem of Hopf Modules which says that a right Hopf module over H is isomorphic to a trivial right Hopf module. We prove a special case of the Fundamental Theorem: if H is a finite dimensional Hopf algebra and $\int_{H*}^l$ is the space of left integrals of H^*, then $H^* \cong \int_{H*}^l \otimes H$ as right Hopf modules, where $\int_{H*}^l \otimes H$ has a trivial right Hopf module structure.

Next we consider Hopf algebras over rings. Many of the properties of Hopf algebras over fields carry over to rings, including under certain conditions, the Fundamental Theorem of Hopf Modules. We define an R-Hopf order H in KG, where R is an integrally closed, integral domain with field of fractions K, G is a finite group, and KG is the group ring K-Hopf algebra. In many ways, R-Hopf orders H in KG play the role of fractional ideals over R in K. We show that an R-Hopf order in KG is an R-Hopf algebra with structure maps induced from KG, and give a collection of R-Hopf orders in KC_p, where C_p denotes the cyclic group of order p. Hopf orders in group rings will have a role to play in the generalization of Galois extensions in Chap. 4 (§4.5).

Chapter 4 consists of three applications of Hopf algebras. The first application concerns quasitriangular structures for bialgebras and Hopf algebras. We show that if H is a quasitriangular Hopf algebra, then there is a solution to the Quantum Yang–Baxter Equation (QYBE). We then introduce an infinite group $\mathcal{B}$ called the braid group on three strands, whose defining property is the braid relation. A solution to the QYBE translates to quantities that satisfy the braid relation, and consequently, an n-dimensional quasitriangular Hopf algebra H determines an n^3-dimensional representation of the braid group $\rho : \mathcal{B} \to \mathrm{GL}_{n^3}(K)$.

The second application relates affine varieties and Hopf algebras. We define affine varieties Λ over a field K and their coordinate rings $K[\Lambda]$ and give some examples. The Hilbert Basis Theorem is applied to show that an affine variety Λ can be identified with the collection of K-algebra homomorphisms $\mathrm{Hom}_{\mathrm{K-alg}}(K[\Lambda], K)$, thus we can think of the variety algebraically, through its coordinate ring. When the coordinate ring is a bialgebra we get a monoid structure on the points of the variety; if the coordinate ring is a Hopf algebra, then there is a group structure on the points of the variety. In this way we give an algebraic structure to a geometric object.

In the third application, we show how Hopf algebras can be used to generalize the notion of a Galois extension. A Galois extension L/K with group G is equivalent to the notion that L is a Galois KG-extension of K where the KG action on L is induced from the classical Galois action of G on L. It is in this latter form (L is a Galois KG-extension of K) that the concept of Galois extension can be extended to rings of integers—we give necessary and sufficient conditions for the ring of integers S of L to be a Galois RG-extension of R.

Significantly, the concept of Galois KG-extension can be generalized to arbitrary K-Hopf algebras (other than the group ring KG) and to other actions (other than the classical action of G as the Galois group). For instance, we show that the splitting field $L/\mathbb{Q}$ of the polynomial $x^3 - 2$ is a Galois $\mathbb{Q}S_3$-extension of $\mathbb{Q}$ with the classical Galois action of S_3 on L, as well as a Galois H-extension of $\mathbb{Q}$ in which H is some other $\mathbb{Q}$-Hopf algebra not the group ring $\mathbb{Q}S_3$, whose action on L is different from the classical Galois group action of S_3. Moreover, in the cyclic order p case, $G = C_p$, we find a Hopf order H in KC_p and a Galois extensions L/K with group C_p whose ring of integers S is a Galois H-extension of R where the Galois action of H on S is the classical Galois action of C_p on L.

Chapters 1–4 begin with a chapter overview which provides a road map for the reader showing what material will be covered, and each section begins with a brief outline of its contents. At the end of each chapter, we collect exercises which review and reinforce the material in the corresponding sections. These exercises range from straightforward applications of the theory to problems designed to challenge the reader. Occasionally, we include a list of "Questions for Further Study" which pose problems suitable for master's degree research projects.

The idea for this book arose as a precursor to the author's book *An Introduction to Hopf Algebras* (Springer, 2011), which treats commutative and cocommutative Hopf algebras over commutative rings with unity. *An Introduction* shows how these Hopf algebras arise as the representing algebras A of representable group functors $F = \mathrm{Hom}_{R\text{-alg}}(A, -)$ on the category of commutative algebras over a commutative ring with unity R. The key result is that if A is a commutative R-algebra, then $A \otimes A$ is the coproduct in the category of commutative R-algebras, and so, if A represents F, then $A \otimes A$ represents $F \times F$. Consequently, by Yoneda's Lemma an algebra map $\Delta : A \to A \otimes A$ (comultiplication) corresponds to a binary operation $F \times F \to F$. If the binary operation admits an identity element and inverses (again given through algebra maps on A satisfying certain conditions), then F is a group and A is an R-Hopf algebra.

However, to put a group structure on $\mathrm{Hom}_{R\text{-alg}}(A, S)$ we only require S to be commutative; the Hopf algebra A can be non-commutative or non-cocommutative (or both: a quantum group). The point is we don't have to work exclusively in the category of commutative algebras. Thus our approach here is broader—the Hopf algebras in this work are developed directly from the notions of algebras, coalgebras, and bialgebras; they are not necessarily commutative or cocommutative.

I owe incalculable thanks and gratitude to the two readers of earlier versions of this book. Their comments, suggestions, and corrections were invaluable to the shaping of the final manuscript.

Unquestionably, without the support of my wife, Rebecca, and my son Andre, I would not have completed this book, and to them I express my greatest thanks and appreciation. I would also like to thank some close friends and colleagues who have helped me over the years: Professors Lindsay Childs, Nigel Byott, Warren Nichols, Timothy Kohl, Alan Koch, Griff Elder, Paul Truman, James Carter, Enoch Lee, Matthew Ragland, Luis Cueva-Parra, and Yi Wang.

Finally, I thank Ann Kostant and Elizabeth Loew at Springer for their support for this book project. From my initial proposal to the final draft, their advice, guidance and encouragement has been critical to the success of this endeavor.

Montgomery, AL, USA   Robert G. Underwood

Contents

Notation

$\mathbb{N}$	Natural numbers $= \{1, 2, 3, \dots\}$
$\mathbb{Z}$	Ring of integers
$\mathbb{Z}^+$	Positive integers
$\mathbb{Q}$	Field of rational numbers
$\mathbb{Q}^\times$	Non-zero rationals
$\mathbb{R}$	Field of real numbers
$\mathbb{R}^+$	Positive real numbers
$\mathbb{R}^\times$	Non-zero real numbers
$\mathbb{C}$	Complex numbers
ζ_m	Primitive mth root of unity
$A \subset B$	A is a proper subset of B
$A \subseteq B$	A is a subset of B
$H < G$	H is a proper subgroup of G
$H \leq G$	H is a subgroup of G
$H \triangleleft G$	H is a normal subgroup of G
$|G|$	Order of the finite group G
$\bar{a}$	Image of a under canonical surjection $A \to A/B$
Z_n	Residue class group modulo n, residue class ring modulo n
D_3	3rd order dihedral group
S_n	Symmetric group on n letters
$\mathcal{V}$	Klein 4-group
C_m	Cyclic group of order m
$\hat{G}$	Character group of G
$[L : K]$	Degree of L over K
K_P	Completion of K at the prime ideal P
R_P	Valuation ring of K_P
π	Uniformizing parameter
$\mathbb{Z}_p$	Ring of p-adic integers
$\mathbb{Q}_p$	Field of p-adic rationals
$\mathrm{disc}(M)$	Discriminant of M

M^D	Dual module
$\mathbb{F}_p$	Finite field with p elements
$\mathrm{GF}(p^n)$	Galois field with p^n elements
V	Vector space
V^*	Dual space
A°	Finite dual
$\int_H^l, \int_H^r$	Space of left, right integrals of H
$H(i)$	One parameter Hopf order
$\mathcal{B}$	Braid group
Λ	Affine variety
$K[\Lambda]$	Coordinate ring of Λ

Chapter 1
Algebras and Coalgebras

In this chapter we introduce algebras and coalgebras. We begin by generalizing the construction of the tensor product to define the tensor product of a finite collection of R-modules, where R is a commutative ring with unity. We specialize to tensor products over a field K and give the diagram-theoretic definition of a K-algebra (A, m_A, λ_A). We then define coalgebras $(C, \Delta_C, \epsilon_c)$ as co-objects to algebras formed by reversing the arrows in the diagrams for algebras.

We next consider the linear dual. We show that if $(C, \Delta_C, \epsilon_C)$ is a coalgebra, then $(C^*, m_{C^*}, \lambda_{C^*})$ is an algebra where the maps m_{C^*} and λ_{C^*} are induced from the transposes of Δ_C and ϵ_C, respectively. The converse of this statement is not true, however: if A is an algebra, then it is not always true that A^* is a coalgebra with structure maps induced from the transposes of the maps for A. The trick is to replace A^* with a certain subspace A° called the finite dual (in fact, A° is the largest subspace of A^* for which $m_A^*(A^\circ) \subseteq A^* \otimes A^*$). Now, if A is an algebra, then A° is a coalgebra. As an application we show that the finite dual $K[x]^\circ$ can be identified with the collection of linearly recursive sequences of all orders over K.

1.1 Multilinear Maps and Tensor Products

In this section we extend the construction of the tensor product $M \otimes_R N$ of R-modules. We generalize R-bilinear maps to R-n-linear maps to define the tensor product of a set of R-modules $M_1, M_2, \ldots, M_n$ as the solution to a universal mapping problem. We show that tensor products can be identified with iterated tensor products in some association.

* * *

Let $n \geq 2$ be an integer, let $M_1, M_2, \ldots, M_n$ be a collection of R-modules, and let A be an R-module.

© Springer International Publishing Switzerland 2015
R.G. Underwood, *Fundamentals of Hopf Algebras*, Universitext,
DOI 10.1007/978-3-319-18991-8_1

Definition 1.1.1. A function $f : M_1 \times M_2 \times \cdots \times M_n \to A$ is R-n-**linear** if for all i, $1 \leq i \leq n$, and all $a_i, a_i' \in M_i, r \in R$,

(i) $f(a_1, a_2, \ldots, a_i + a_i', \ldots, a_n) = f(a_1, a_2, \ldots, a_i, \ldots, a_n) + f(a_1, a_2, \ldots, a_i', \ldots, a_n)$,

(ii) $f(a_1, a_2, \ldots, ra_i, \ldots, a_n) = rf(a_1, a_2, \ldots, a_i, \ldots, a_n)$.

For instance, an R-bilinear map is an R-2-linear map.

Definition 1.1.2. A **tensor product** of $M_1, M_2, \ldots, M_n$ over R is an R-module $M_1 \otimes M_2 \otimes \cdots \otimes M_n$ together with an R-n-linear map

$$f : M_1 \times M_2 \times \cdots \times M_n \to M_1 \otimes M_2 \otimes \cdots \otimes M_n$$

so that for every R-module A and R-n-linear map $h : M_1 \times M_2 \times \cdots \times M_n \to A$ there exists a unique R-module map $\tilde{h} : M_1 \otimes M_2 \otimes \cdots \otimes M_n \to A$ for which $\tilde{h}f = h$, that is, the following diagram commutes.

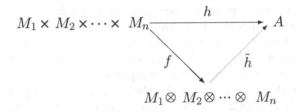

We construct a tensor product as follows. Let $F\langle M_1 \times M_2 \times \cdots \times M_n \rangle$ denote the free R-module on the set $M_1 \times M_2 \times \cdots \times M_n$. Let J be the submodule of $F\langle M_1 \times M_2 \times \cdots \times M_n \rangle$ generated by quantities of the form

$$(a_1, a_2, \ldots, a_i + a_i', \ldots, a_n) - (a_1, a_2, \ldots, a_i, \ldots, a_n) - (a_1, a_2, \ldots, a_i', \ldots, a_n),$$

$$(a_1, a_2, \ldots, ra_i, \ldots, a_n) - r(a_1, a_2, \ldots, a_i, \ldots, a_n),$$

for all i, $1 \leq i \leq n$, and all $a_i, a_i' \in M_i, r \in R$. Let

$$\iota : M_1 \times M_2 \times \cdots \times M_n \to F\langle M_1 \times M_2 \times \cdots \times M_n \rangle$$

be the natural inclusion map and let

$$s : F\langle M_1 \times M_2 \times \cdots \times M_n \rangle \to F\langle M_1 \times M_2 \times \cdots \times M_n \rangle / J$$

be the canonical surjection. Let $f = s\iota$. Then the quotient space $F\langle M_1 \times M_2 \times \cdots \times M_n \rangle / J$ together with the map f (which is clearly R-n-linear) is a tensor product; it solves the universal mapping problem described in Definition 1.1.2.

Proposition 1.1.3. $F\langle M_1 \times M_2 \times \cdots \times M_n \rangle / J$ together with the map f is a tensor product of $M_1, M_2, \ldots, M_n$ over R.

Proof. We show that the conditions of Definition 1.1.2 are satisfied. Let A be an R-module and let $h : M_1 \times M_2 \times \cdots \times M_n \to A$ be an R-n-linear map. There exists an R-module homomorphism

$$\phi : F\langle M_1 \times M_2 \times \cdots \times M_n \rangle \to A,$$

defined as $\phi(a_1, a_2, \ldots, a_n) = h(a_1, a_2, \ldots, a_n)$, $\forall (a_1, a_2, \ldots, a_n) \in M_1 \times M_2 \times \cdots \times M_n$. Since ϕ is R-n-linear, $J \subseteq \ker(\phi)$, and so, by the universal mapping property for kernels, there exists an R-module homomorphism

$$\tilde{h} : F\langle M_1 \times M_2 \times \cdots \times M_n \rangle / J \to A,$$

defined as $\tilde{h}((a_1, a_2, \ldots, a_n) + J) = \phi(a_1, a_2, \ldots, a_n) = h(a_1, a_2, \ldots, a_n)$. Now for all $(a_1, a_2, \ldots, a_n) \in M_1 \times M_2 \times \cdots \times M_n$, $\tilde{h}\iota(a_1, a_2, \ldots, a_n) = h(a_1, a_2, \ldots, a_n)$, and so, $\tilde{h}f = h$. Moreover, $\tilde{h}$ is unique because

$$\{(a_1, a_2, \ldots, a_n) + J : (a_1, a_2, \ldots, a_n) \in M_1 \times M_2 \times \cdots \times M_n\}$$

is a set of generators for $F\langle M_1 \times M_2 \times \cdots \times M_n \rangle / J$.   □

As a consequence of Proposition 1.1.3, we write

$$F\langle M_1 \times M_2 \times \cdots \times M_n \rangle / J = M_1 \otimes M_2 \otimes \cdots \otimes M_n,$$

with the coset $(a_1, a_2, \ldots, a_n) + J$ now written as the **tensor** $a_1 \otimes a_2 \otimes \cdots \otimes a_n$.

Proposition 1.1.4. *Let M_1, M_2 be R-modules and let N_1 be an R-submodule of M_1 and let N_2 be an R-submodule of M_2. Then there is an isomorphism of R-modules*

$$M_1/N_1 \otimes_R M_2/N_2 \cong (M_1 \otimes_R M_2)/(N_1 \otimes_R M_2 + M_1 \otimes_R N_2).$$

Proof. First note that there is an R-bilinear map

$$h : M_1 \times M_2 \to M_1/N_1 \otimes_R M_2/N_2,$$

defined by $h(a, b) = (a + N_1) \otimes (b + N_2)$. Since $M_1 \otimes_R M_2$ is a tensor product there exists an R-module map

$$\tilde{h} : M_1 \otimes_R M_2 \to M_1/N_1 \otimes_R M_2/N_2,$$

defined as $\tilde{h}(a \otimes b) = (a + N_1) \otimes (b + N_2)$. Now $N_1 \otimes_R M_2 + M_1 \otimes_R N_2 \subseteq \ker(\tilde{h})$, and so by the universal mapping property for kernels, there exists an R-module map

$$\alpha : (M_1 \otimes_R M_2)/(N_1 \otimes_R M_2 + M_1 \otimes_R N_2) \to M_1/N_1 \otimes_R M_2/N_2,$$

with $\alpha(a \otimes b + (N_1 \otimes_R M_2 + M_1 \otimes_R N_2)) = (a + N_1) \otimes (b + N_2)$.

Next, let

$$l : M_1/N_1 \times M_2/N_2 \to (M_1 \otimes_R M_2)/(N_1 \otimes_R M_2 + M_1 \otimes_R N_2)$$

be the relation defined as

$$l(a + N_1, b + N_2) = a \otimes b + (N_1 \otimes_R M_2 + M_1 \otimes_R N_2),$$

for $a \in M_1$, $b \in M_2$. We claim that l is actually a function on $M_1/N_1 \times M_2/N_2$. To this end, let $x = a + n$, $y = b + n'$ for some $n \in N_1$, $n' \in N_2$. Then

$$
\begin{aligned}
l(x + N_1, y + N_2) &= x \otimes y + (N_1 \otimes_R M_2 + M_1 \otimes_R N_2) \\
&= (a + n) \otimes (b + n') + (N_1 \otimes_R M_2 + M_1 \otimes_R N_2) \\
&= a \otimes b + a \otimes n' + n \otimes b + n \otimes n' + (N_1 \otimes_R M_2 + M_1 \otimes_R N_2),
\end{aligned}
$$

and thus $l(a + N_1, b + N_2) - l(x + N_1, y + N_2) = N_1 \otimes_R M_2 + M_1 \otimes_R N_2$. It follows that l is a well-defined function on $M_1/N_1 \times M_2/N_2$. As one can easily check, l is R-bilinear and since $M_1/N_1 \otimes_R M_2/N_2$ is a tensor product, there exists an R-module map

$$\tilde{l} : M_1/N_1 \otimes_R M_2/N_2 \to (M_1 \otimes_R M_2)/(N_1 \otimes_R M_2 + M_1 \otimes_R N_2),$$

defined as $\tilde{l}((a + N_1) \otimes (b + N_2)) = a \otimes b + (N_1 \otimes_R M_2 + M_1 \otimes_R N_2)$. Clearly $\alpha^{-1} = \tilde{l}$, and thus $\tilde{l}$ is an isomorphism. $\square$

Let M_1, M_2, M_3 be R-modules. Then the associative property for tensor products holds.

Proposition 1.1.5. *There is an R-module isomorphism $M_1 \otimes (M_2 \otimes M_3) \cong (M_1 \otimes M_2) \otimes M_3$.*

Proof. Let $h : M_1 \times M_2 \times M_3 \to (M_1 \otimes M_2) \otimes M_3$ be the map defined by $(a_1, a_2, a_3) \mapsto (a_1 \otimes a_2) \otimes a_3$. Then

$$
\begin{aligned}
h(a_1 + a_1', a_2, a_3) &= ((a_1 + a_1') \otimes a_2) \otimes a_3 \\
&= (a_1 \otimes a_2 + a_1' \otimes a_2) \otimes a_3 \\
&= (a_1 \otimes a_2) \otimes a_3 + (a_1' \otimes a_2) \otimes a_3 \\
&= h(a_1, a_2, a_3) + h(a_1', a_2, a_3),
\end{aligned}
$$

and so, h is linear in the first component. Likewise h is linear in the other components. Also,

$$h(ra_1, a_2, a_3) = (ra_1 \otimes a_2) \otimes a_3$$
$$= r(a_1 \otimes a_2) \otimes a_3$$
$$= r((a_1 \otimes a_2) \otimes a_3)$$
$$= rh(a_1, a_2, a_3),$$

and so h is R-linear in the first component. Likewise, h is an R-linear in the other components. Thus h is an R-3-linear function. Since $M_1 \otimes M_2 \otimes M_3$ is a tensor product over R, there exists a unique map of R-modules

$$\tilde{h} : M_1 \otimes M_2 \otimes M_3 \to (M_1 \otimes M_2) \otimes M_3$$

defined by $a_1 \otimes a_2 \otimes a_3 \mapsto (a_1 \otimes a_2) \otimes a_3$. Clearly $\tilde{h}$ is an isomorphism.

In a similar manner one constructs an isomorphism

$$\tilde{g} : M_1 \otimes M_2 \otimes M_3 \to M_1 \otimes (M_2 \otimes M_3)$$

defined by $a_1 \otimes a_2 \otimes a_3 \mapsto a_1 \otimes (a_2 \otimes a_3)$. Let

$$\phi : M_1 \otimes (M_2 \otimes M_3) \cong (M_1 \otimes M_2) \otimes M_3$$

be the composition $\phi = \tilde{h} \circ \tilde{g}^{-1}$ given by $a_1 \otimes (a_2 \otimes a_3) \mapsto (a_1 \otimes a_2) \otimes a_3$. Then ϕ is an isomorphism of R-modules. $\square$

By an "iterated tensor product in some association" we mean a tensor product whose factors themselves may be tensor products or tensor products of tensor products, and so on. For example, for the R-modules M_1, M_2, M_3

$$M_1 \otimes (M_2 \otimes M_3) \text{ and } (M_1 \otimes M_2) \otimes M_3$$

are iterated tensor products in some association. For the R-modules $M_1, M_2, M_3,$ M_4, M_5

$$(M_1 \otimes M_2) \otimes (M_3 \otimes M_4 \otimes M_5) \text{ and } \quad M_1 \otimes ((M_2 \otimes M_3) \otimes M_4) \otimes M_5$$

are iterated tensor products in some association. There is a "natural" isomorphism between tensor products and iterated tensor products in some association. By natural we mean that the map is defined without using any of the properties of the tensor product—we just add parentheses. For example, there is a natural isomorphism

$$M_1 \otimes M_2 \otimes M_3 \to (M_1 \otimes M_2) \otimes M_3,$$

defined by $a_1 \otimes a_2 \otimes a_3 \mapsto (a_1 \otimes a_2) \otimes a_3$ and a natural isomorphism

$$M_1 \otimes M_2 \otimes M_3 \otimes M_4 \otimes M_5 \to M_1 \otimes ((M_2 \otimes M_3) \otimes M_4) \otimes M_5$$

defined by

$$a_1 \otimes a_2 \otimes a_3 \otimes a_4 \otimes a_5 \mapsto a_1 \otimes ((a_2 \otimes a_3) \otimes a_4) \otimes a_5.$$

Proposition 1.1.6. *Let $M_1, M_2, \ldots, M_n$ be R-modules and let S be an iterated tensor product of $M_1, M_2, \ldots, M_n$ in some association. Then there is a natural isomorphism $M_1 \otimes M_2 \otimes \cdots \otimes M_n \cong S$.*

Proof. We proceed by induction on n. The trivial case $n = 2$ clearly holds. Assume the result holds for any collection of R-modules with $M_1, M_2, \ldots, M_s$ with $2 \le s < n$. There exists an integer r, $2 \le r < n$ for which $S = T \otimes U$ where T is an iterated tensor product of $M_1, M_2, \ldots, M_r$ in some association, and U is an iterated tensor product of $M_{r+1}, M_{r+2}, \ldots, M_n$ in some association. By the induction hypothesis,

$$S \cong (M_1 \otimes M_2 \otimes \cdots \otimes M_r) \otimes (M_{r+1} \otimes M_{r+2} \otimes \cdots \otimes M_n),$$

and so, $S \cong M_1 \otimes M_2 \otimes \cdots \otimes M_n$, see §1.4, Exercise 2. $\square$

   In view of Proposition 1.1.6 we will "ignore the parentheses" and consider tensor products and iterated tensor products with some association as the same objects through the natural isomorphism.
   We close this section with a few remarks about maps.

Proposition 1.1.7. *Let $M_1, M_2, \ldots, M_n, M_1', M_2', \ldots, M_n'$, be R-modules and let $f_i :$ $M_i \to M_i'$, for $1 \le i \le n$, be R-module maps. There exists a unique map of R-modules*

$$(f_1 \otimes f_2 \otimes \cdots \otimes f_n) : M_1 \otimes M_2 \otimes \cdots \otimes M_n \to M_1' \otimes M_2' \otimes \cdots \otimes M_n'$$

defined as

$$(f_1 \otimes f_2 \otimes \cdots \otimes f_n)(a_1 \otimes a_2 \otimes \cdots \otimes a_n) = f_1(a_1) \otimes f_2(a_2) \otimes \cdots \otimes f_n(a_n)$$

for all $a_i \in M_i$.

Proof. There exists an R-n-linear map

$$(f_1 \times f_2 \times \cdots \times f_n) : M_1 \times M_2 \times \cdots \times M_n \to M_1' \otimes M_2' \otimes \cdots \otimes M_n'$$

defined as

$$(f_1 \times f_2 \times \cdots \times f_n)(a_1, a_2, \ldots, a_n) = f_1(a_1) \otimes f_2(a_2) \otimes \cdots \otimes f_n(a_n)$$

for all $a_i \in M_i$. Now use the fact that $M_1 \otimes M_2 \otimes \cdots \otimes M_n$ is a tensor product to induce the map

$$\tilde{h} : M_1 \otimes M_2 \otimes \cdots \otimes M_n \to M_1' \otimes M_2' \otimes \cdots \otimes M_n',$$

defined as

$$\tilde{h}(a_1 \otimes a_2 \otimes \cdots \otimes a_n) = f_1(a_1) \otimes f_2(a_2) \otimes \cdots \otimes f_n(a_n)$$

for all $a_i \in M_i$. Set $f_1 \otimes f_2 \otimes \cdots \otimes f_n = \tilde{h}$. $\qquad\square$

As an illustration of Proposition 1.1.7, take $n = 2$ with $M_1 = M \otimes M$, $M_2 = M$, $f_1 : M \otimes M \to M_1'$, $f_2 : M \to M_2'$. Then there is an R-linear map

$$f_1 \otimes f_2 : (M \otimes M) \otimes M \to M_1' \otimes M_2'$$

defined as

$$(f_1 \otimes f_2)((a \otimes b) \otimes c) = f_1(a \otimes b) \otimes f_2(c).$$

But in view of our convention to ignore parentheses, this is the map

$$f_1 \otimes f_2 : M \otimes M \otimes M \to M_1' \otimes M_2'$$

defined as

$$(f_1 \otimes f_2)(a \otimes b \otimes c) = f_1(a \otimes b) \otimes f_2(c).$$

Corollary 1.1.8. *Let K be a field and let V_i, $1 \le i \le n$, be a finite set of vector spaces over K. Then*

$$V_1^* \otimes V_2^* \otimes \cdots \otimes V_n^* \subseteq (V_1 \otimes V_2 \otimes \cdots \otimes V_n)^*.$$

Proof. Let $f_i : V_i \to K$, $1 \le i \le n$, be a set of K-module maps (that is, $f_i \in V_i^*$, $\forall i$). By Proposition 1.1.7 (with $V_i' = K$), there exists a unique K-module map

$$(f_1 \otimes f_2 \otimes \cdots \otimes f_n)(a_1 \otimes a_2 \otimes \cdots \otimes a_n) = f_1(a_1) \otimes f_2(a_2) \otimes \cdots \otimes f_n(a_n)$$

Now,

$$f_1(a_1) \otimes f_2(a_2) \otimes \cdots \otimes f_n(a_n) = f_1(a_1)f_2(a_2) \cdots f_n(a_n) \in K,$$

since $K \otimes_K K \cong K$ through the map $r \otimes s \mapsto rs$. Consequently,

$$V_1^* \otimes V_2^* \otimes \cdots \otimes V_n^* \subseteq (V_1 \otimes V_2 \otimes \cdots \otimes V_n)^*. \qquad \square$$

We remark that we have equality in Corollary 1.1.8 if and only if each V_i is finite dimensional.

1.2 Algebras and Coalgebras

In this section we present the diagram-theoretic definition of a K-algebra (A, m_A, λ_A) and compare this definition to the algebras that most readers already know. We discuss quotient algebras and algebra homomorphisms. Next, we define coalgebras $(C, \Delta_C, \epsilon_c)$ as co-objects to algebras formed by reversing the arrows in the diagrams for the algebras and give some basic examples. We introduce notation due to M. Sweedler, "Sweedler notation" to write the image of the comultiplication map and we show how Sweedler notation works to simplify computations. We define coideals, quotient coalgebras, and coalgebra homomorphisms.

$$* \quad * \quad *$$

Let K be a field.

Definition 1.2.1. A **K-algebra** is a triple (A, m_A, λ_A) consisting of a vector space A over K and K-linear maps $m_A : A \otimes_K A \to A$ and $\lambda_A : K \to A$ that satisfy the following conditions.

(i) The diagram commutes:

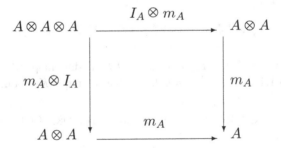

Here the map $I_A : A \to A$ is the identity map and the maps $I_A \otimes m_A : A \otimes A \otimes A \to A \otimes A$ and $m_A \otimes I_A : A \otimes A \otimes A \to A \otimes A$ are defined by $a \otimes b \otimes c \mapsto a \otimes m_A(b \otimes c)$ and $a \otimes b \otimes c \mapsto m_A(a \otimes b) \otimes c$, for all $a, b, c \in A$, respectively. Equivalently, we have for all $a, b, c \in A$

$$m_A(I_A \otimes m_A)(a \otimes b \otimes c) = m_A(m_A \otimes I_A)(a \otimes b \otimes c) \qquad (1.1)$$

(ii) The diagrams commute:

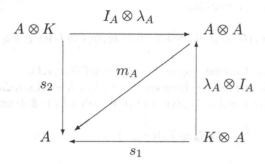

Here the maps $I_A \otimes \lambda_A$, $\lambda_A \otimes I_A$ are defined by $a \otimes r \mapsto a \otimes \lambda_A(r)$, $r \otimes a \mapsto \lambda_A(r) \otimes a$ for all $r \in K$, $a \in A$, respectively. The map $s_1 : K \otimes A \to A$ is defined by $r \otimes a \mapsto ra$ and the map $s_2 : A \otimes K \to A$ is defined as $a \otimes r \mapsto ra$. Equivalently, we have for all $r \in K$, $a \in A$,

$$m_A(I_A \otimes \lambda_A)(a \otimes r) = ra = m_A(\lambda_A \otimes I_A)(r \otimes a). \qquad (1.2)$$

The map m_A is the **multiplication map** and the map λ_A is the **unit map**. Property (1.1) is the **associative property** and property (1.2) is the **unit property**.

Here is the usual definition of K-algebra.

Definition 1.2.1′. A **K-algebra** is a ring A with unity 1_A together with a ring homomorphism $\lambda_A : K \to A$ which satisfies $\lambda_A(r)a = a\lambda_A(r)$ for $a \in A$, $r \in R$. Then A is a vector space over K with scalar multiplication given as

$$ra = \lambda_A(r)a = a\lambda_A(r) \qquad (1.3)$$

for $r \in K$, $a \in A$.

Our first task is to show that we really don't have a new definition of K-algebra.

Proposition 1.2.2. *K-algebras as defined by Definition 1.2.1′ coincide with K-algebras as defined by Definition 1.2.1.*

Proof. Let A be a K-algebra as in Definition 1.2.1′. Let $B : A \times A \to A$ be the map defined as $(a, b) \mapsto ab$, where ab is the multiplication in the ring A. Now B is a K-bilinear (this follows from the associative and distributive properties of multiplication in A). Since $A \otimes_K A$ is a tensor product, there exists a unique K-linear map $m_A : A \otimes_K A \to A$ defined as

$$m_A\left(\sum a \otimes b\right) = \sum B(a, b) = \sum ab.$$

Let $I_A : A \to A$ denote the identity map. The associative property of the multiplication in A implies that

$$m_A(I_A \otimes m_A)(a \otimes b \otimes c) = m_A(m_A \otimes I_A)(a \otimes b \otimes c)$$

for all $a, b, c \in A$. Thus the map m_A satisfies condition (1.1).

We next show that the ring homomorphism $\lambda_A : K \to A$ satisfies condition (1.2). For $r, s \in K$, $\lambda_A(rs) = \lambda_A(r)\lambda_A(s) = r\lambda_A(s)$, so that λ_A is K-linear. From (1.3),

$$m_A(I_A \otimes \lambda_A)(a \otimes r) = a\lambda_A(r) = ra,$$

and

$$m_A(\lambda_A \otimes I_A)(r \otimes a) = \lambda_A(r)a = ra,$$

which shows that (1.2) holds. Thus the triple (A, m_A, λ_A) is a K-algebra.

Conversely, suppose that (A, m_A, λ_A) is a K-algebra as in Definition 1.2.1. Define multiplication on A as $ab = m_A(a \otimes b)$, for $a, b \in A$. Then A is a ring. From (1.2) for $r \in K$, $a \in A$,

$$ra = m_A(I_A \otimes \lambda_A)(a \otimes r) = m_A(a \otimes \lambda_A(r)) = a\lambda_A(r),$$

and

$$ra = m_A(\lambda_A \otimes I_A)(r \otimes a) = m_A(\lambda_A(r) \otimes a) = \lambda_A(r)a,$$

thus $\lambda_A(K)$ is in the center of A. Setting $r = 1_K$, $1_A = \lambda_A(1_K)$, shows that A is a ring with unity, 1_A. Setting $a = \lambda_A(s)$ for $s \in K$ shows that λ_A is a ring homomorphism. Clearly the scalar multiplication on A is defined through λ_A, and so, A is a K-algebra in the sense of Definition 1.2.1'. $\square$

To simplify notation we will usually write the K-algebra (A, m_A, λ_A) as A.

The K-algebra A is **commutative** if

$$m_A \tau = m_A,$$

where τ denotes the **twist map** defined as $\tau(a \otimes b) = b \otimes a$ for $a, b \in A$.

Example 1.2.3. The field K is an algebra over itself with $m_K : K \otimes_K K \to K$ defined as $r \otimes s \mapsto rs$ and $\lambda_K : K \to K$ given as $r \mapsto r$ for all $r, s \in K$.

Example 1.2.4. The polynomial ring $K[x]$ is a K-algebra with $m_{K[x]} : K[x] \otimes_K K[x] \to K[x]$ given by ordinary polynomial multiplication and $\lambda_{K[x]} : K \to K[x]$ defined as $r \mapsto r1$, for all $r \in K$.

Example 1.2.5. Let G be any finite group with identity element 1. The group ring KG is a K-algebra with $m_{KG} : KG \otimes_K KG \to KG$ defined by $g \otimes h \mapsto gh$ and $\lambda_{KG} : K \to KG$ given as $r \mapsto r1$, for all $g, h \in G$, $r \in K$. Since K is a field, the image

$\lambda_{KG}(K)$ is isomorphic to K; KG contains a copy of K through the identification $r = r1$. The unit map is then given as $\lambda_{KG}(r) = r$.

Example 1.2.6. Let $L = K(\alpha)$ be a simple algebraic extension of K. Then L is a K-algebra with $m_L : L \otimes_K L \to L$ given by multiplication in the field L and $\lambda_L : K \to L$ defined as $r \mapsto r$, for all $r \in K$.

Clearly, the K-algebras of Examples 1.2.3, 1.2.4, and 1.2.6 are commutative, while KG is a commutative K-algebra if and only if G is abelian.

Let A, B be K-algebras. The tensor product $A \otimes B$ has the structure of a K-algebra with multiplication

$$m_{A \otimes B} : (A \otimes B) \otimes (A \otimes B) \to A \otimes B$$

defined by

$$
\begin{aligned}
m_{A \otimes B}((a \otimes b) \otimes (c \otimes d)) &= (m_A \otimes m_B)(I_A \otimes \tau \otimes I_B)(a \otimes (b \otimes c) \otimes d) \\
&= (m_A \otimes m_B)(a \otimes (c \otimes b) \otimes d) \\
&= (m_A \otimes m_B)((a \otimes c) \otimes (b \otimes d)) \\
&= ac \otimes bd
\end{aligned}
$$

for $a, c \in A$, $b, d \in B$. The unit map $\lambda_{A \otimes B} : K \to A \otimes B$ is given as

$$\lambda_{A \otimes B}(r) = \lambda_A(r) \otimes 1_B$$

for $r \in K$.

A K-algebra A is a ring with addition given by vector addition and multiplication defined by m_A.

Proposition 1.2.7. *Let A be a K-algebra and let I be an ideal of A. Then the quotient space A/I is a K-algebra.*

Proof. We need to define a multiplication map $m_{A/I}$ and a unit map $\lambda_{A/I}$. Let $s : A \to A/I$ denote the canonical quotient map. The composition

$$s \circ m_A : A \otimes A \to A/I$$

is a map of K-vector spaces defined as $(s \circ m_A)(a \otimes b) = ab + I$. Note that $I \otimes A + A \otimes I$ is a subspace of $A \otimes A$. Let $a \otimes b + c \otimes d \in I \otimes A + A \otimes I$ for $a, d \in I$, $b, c \in A$. Since I is an ideal, $m_A(a \otimes b + c \otimes d) = ab + cd \in I$, hence $I \otimes A + A \otimes I \subseteq \ker(s \circ m_A)$. Thus by the universal mapping property for kernels, there is a map of vector spaces

$$\overline{s \circ m_A} : (A \otimes A)/(I \otimes A + A \otimes I) \to A/I$$

defined as

$$\overline{s \circ m_A}(a \otimes b + (I \otimes A + A \otimes I)) = ab + I.$$

By Proposition 1.1.4 there is an isomorphism of vector spaces

$$\tilde{\beta} : A/I \otimes A/I \to (A \otimes A)/(I \otimes A + A \otimes I)$$

given as

$$\tilde{\beta}((a + I) \otimes (b + I)) = a \otimes b + (I \otimes A + A \otimes I).$$

Let $m_{A/I}$ denote the composition $\overline{s \circ m_A} \circ \tilde{\beta}$. The map

$$m_{A/I} : A/I \otimes A/I \to A/I,$$

given as $(a + I) \otimes (b + I) \mapsto ab + I$, now serves as the multiplication map of A/I. As one can check, $m_{A/I}$ satisfies the associative property since m_A does.

For the unit map of A/I, let $\lambda_{A/I}$ be the composition $s \circ \lambda_A : K \to A/I$. Then it is easily checked that $\lambda_{A/I}$ satisfies the unit property. Thus $(A/I, m_{A/I}, \lambda_{A/I})$ is a K-algebra.   □

The K-algebra A/I is the **quotient algebra of A by I**.

Let (A, m_A, λ_A), (B, m_B, λ_B) be K-algebras. A K**-algebra homomorphism** from A to B is a map of additive groups $\phi : A \to B$ (that is, $\phi(a + b) = \phi(a) + \phi(b)$, for $a, b \in A$) for which $\phi(1_A) = 1_B$

$$\phi(m_A(a \otimes b)) = m_B(\phi(a) \otimes \phi(b)),$$

and $\phi(\lambda_A(r)) = \lambda_B(r)$ for $a, b \in A$, $r \in K$. In particular, for A to be a subalgebra of B (when ϕ is an inclusion) we require that $1_A = 1_B$.

We now describe objects that are dual (in some sense) to algebras; essentially forming them by reversing the arrows in the structure maps for algebras. These objects are called "coalgebras."

Let C be a K-vector space. The scalar multiplication of C defines two maps $s_1 : K \otimes C \to C$ with $r \otimes c \mapsto rc$ and $s_2 : C \otimes K \to C$ with $c \otimes r \mapsto rc$, for $c \in C$, $r \in K$.

Definition 1.2.8. A K**-coalgebra** is a triple $(C, \Delta_C, \epsilon_C)$ consisting of a vector space C over K and K-linear maps $\Delta_C : C \to C \otimes_K C$ and $\epsilon_C : C \to K$ that satisfy the following conditions.

(i) The diagram commutes:

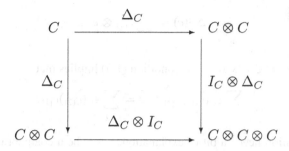

Here the map $I_C : C \to C$ is the identity map and the maps $I_C \otimes \Delta_C : C \otimes C \to C \otimes C \otimes C$ and $\Delta_C \otimes I_C : C \otimes C \to C \otimes C \otimes C$ are defined by $a \otimes b \mapsto a \otimes \Delta_C(b)$ and $a \otimes b \mapsto \Delta_C(a) \otimes b$, for all $a, b \in C$, respectively. Equivalently, we have for all $c \in C$

$$(I_C \otimes \Delta_C)\Delta_C(c) = (\Delta_C \otimes I_C)\Delta_C(c) \tag{1.4}$$

(ii) The diagrams commute:

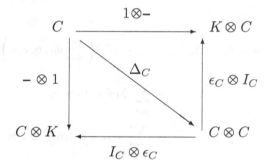

Here the maps $- \otimes 1$ and $1 \otimes -$ are defined by $c \mapsto c \otimes 1$ and $c \mapsto 1 \otimes c$, respectively. Equivalently,

$$(\epsilon_C \otimes I_C)\Delta_C(c) = 1 \otimes c, \quad (I_C \otimes \epsilon_C)\Delta_C(c) = c \otimes 1 \tag{1.5}$$

for all $c \in C$. The maps Δ_C and ϵ_C are the **comultiplication** and **counit** maps, respectively, of the coalgebra C. Condition (1.4) is the **coassociative property** and Condition (1.5) is the **counit property**.

A K-coalgebra C is **cocommutative** if

$$\tau(\Delta_C(c)) = \Delta_C(c),$$

for all $c \in C$.

We use the notation of Sweedler [Sw69, §1.2] to write

$$\Delta_C(c) = \sum_{(c)} c_{(1)} \otimes c_{(2)}.$$

Since $s_1(1 \otimes c) = c = s_2(c \otimes 1)$, condition (1.5) implies that

$$\sum_{(c)} \epsilon_C(c_{(1)})c_{(2)} = c = \sum_{(c)} \epsilon_C(c_{(2)})c_{(1)}. \tag{1.6}$$

Sweedler notation needs a bit of explanation. Let C be a coalgebra and let $c \in C$. Then

$$(I_C \otimes \Delta_C)\Delta_C(c) = (I_C \otimes \Delta_C)\left(\sum_{(c)} c_{(1)} \otimes c_{(2)} \right)$$

$$= \sum_{(c)} c_{(1)} \otimes \Delta_C(c_{(2)})$$

$$= \sum_{(c,c_{(2)})} c_{(1)} \otimes c_{(2)(1)} \otimes c_{(2)(2)}, \tag{1.7}$$

and

$$(\Delta_C \otimes I_C)\Delta_C(c) = (\Delta_C \otimes I_C)\left(\sum_{(c)} c_{(1)} \otimes c_{(2)} \right)$$

$$= \sum_{(c)} \Delta_C(c_{(1)}) \otimes c_{(2)}$$

$$= \sum_{(c,c_{(1)})} c_{(1)(1)} \otimes c_{(1)(2)} \otimes c_{(2)}. \tag{1.8}$$

By (1.4), $(I_C \otimes \Delta_C)\Delta_C = (\Delta_C \otimes I_C)\Delta_C$ and so, the expressions in (1.7) and (1.8) are equal. The common value in (1.7) and (1.8) is denoted as

$$\sum_{(c)} c_{(1)} \otimes c_{(2)} \otimes c_{(3)}.$$

Similarly, the common value of

$$(I_C \otimes I_C \otimes \Delta_C)(I_C \otimes \Delta_C)\Delta_C(c) = (I_C \otimes I_C \otimes \Delta_C)(\Delta_C \otimes I_C)\Delta_C(c)$$

$$= (I_C \otimes \Delta_C \otimes I_C)(\Delta_C \otimes I_C)\Delta_C(c)$$

$$= (I_C \otimes \Delta_C \otimes I_C)(I_C \otimes \Delta_C)\Delta_C(c)$$

$$= (\Delta_C \otimes I_C \otimes I_C)(I_C \otimes \Delta_C)\Delta_C(c)$$

$$= (\Delta_C \otimes I_C \otimes I_C)(\Delta_C \otimes I_C)\Delta_C(c)$$

is denoted as

$$\sum_{(c)} c_{(1)} \otimes c_{(2)} \otimes c_{(3)} \otimes c_{(4)}.$$

The coassociative property gives a well-defined meaning to the coproduct $\sum_{(c)} c_{(1)} \otimes c_{(2)} \otimes c_{(3)}$, just as the associative property in an algebra gives an unambiguous meaning to the product abc since this product is the common value of $a(bc)$ and $(ab)c$.

Here are some examples of coalgebras.

Example 1.2.9. The field K as a vector space over itself is a K-coalgebra where the comultiplication map $\Delta_K : K \to K \otimes K$ is defined as $\Delta_K(a) = a \otimes 1$ and the counit map $\epsilon_K : K \to K$ is given as $\epsilon_K(a) = a$. The K-coalgebra K is the **trivial coalgebra**.

Example 1.2.10. Let x be an indeterminate and let $C = K \oplus Kx$ be the direct sum of vector spaces. Then C is a K-coalgebra with $\Delta_C : C \to C \otimes_K C$ defined on the K-basis $\{1, x\}$ as

$$\Delta_C(1) = 1 \otimes 1, \quad \Delta_C(x) = x \otimes x,$$

$$\epsilon_C(1) = \epsilon_C(x) = 1.$$

Example 1.2.11. Let x be an indeterminate and let $C = K \oplus Kx$ be the direct sum of vector spaces. Then C is a K-coalgebra with $\Delta_C : C \to C \otimes_K C$ defined on the K-basis $\{1, x\}$ as

$$\Delta_C(1) = 1 \otimes 1, \quad \Delta_C(x) = 1 \otimes x + x \otimes 1,$$

$$\epsilon_C(1) = 1, \quad \epsilon_C(x) = 0.$$

Example 1.2.12. Let V denote an n-dimensional K-vector space with basis $\mathcal{B} = \{b_1, b_2, \dots, b_n\}$. Let $\Delta_V : V \to V \otimes V$ be the K-linear map defined on the basis $\mathcal{B}$ as

$$\Delta_V(b_i) = b_i \otimes b_i,$$

and let $\epsilon_V : V \to K$ be the K-linear map defined on $\mathcal{B}$ as

$$\epsilon_V(b_i) = 1.$$

Then as one can easily verify, the triple $(V, \Delta_V, \epsilon_V)$ is a K-coalgebra.

Example 1.2.13. Let $K[x]$ denote the K-vector space of polynomials in the indeterminate x. Let $\Delta_{K[x]} : K[x] \to K[x] \otimes K[x]$ be the K-linear map defined on the K-basis $\{1, x, x^2, \dots\}$ as

$$\Delta_{K[x]}(x^m) = x^m \otimes x^m,$$

and let $\epsilon_{K[x]} : K[x] \to K$ be the K-linear map defined on $\{1, x, x^2, \dots\}$ as

$$\epsilon_{K[x]}(x^m) = 1.$$

Then the triple $(K[x], \Delta_{K[x]}, \epsilon_{K[x]})$ is a K-coalgebra.

Example 1.2.14. Let $K[x]$ denote the K-vector space of polynomials in the indeterminate x. Let $\Delta_{K[x]} : K[x] \to K[x] \otimes K[x]$ be the K-linear map defined on the K-basis $\{1, x, x^2, \dots\}$ as

$$\Delta_{K[x]}(x^m) = \sum_{i=0}^{m} \binom{m}{i} x^i \otimes x^{m-i},$$

and let $\epsilon_{K[x]} : K[x] \to K$ be the K-linear map defined on $\{1, x, x^2, \dots\}$ as

$$\epsilon_{K[x]}(x^m) = \delta_{0,m}.$$

Then the triple $(K[x], \Delta_{K[x]}, \epsilon_{K[x]})$ is a K-coalgebra. This is the **divided power coalgebra**.

We next discuss the analog of an ideal in a ring for a coalgebra. Let C be a K-coalgebra. A subspace $I \subseteq C$ is a **coideal** of C if

$$\Delta_C(I) \subseteq I \otimes C + C \otimes I$$

and $\epsilon_C(I) = 0$.

Proposition 1.2.15. *Let $I \subseteq C$ be a coideal of C. Then the quotient space C/I is a K-coalgebra.*

Proof. Let $s : C \otimes C \to (C \otimes C)/(I \otimes C + C \otimes I)$ be the canonical surjection, and consider the composition

$$(s \circ \Delta_C) : C \to (C \otimes C)/(I \otimes C + C \otimes I).$$

Since $\Delta_C(I) \subseteq I \otimes C + C \otimes I$, $I \subseteq \ker(s \circ \Delta_C)$. Thus by the universal mapping property for kernels, there is a map

$$\overline{s \circ \Delta_C} : C/I \to (C \otimes C)/(I \otimes C + C \otimes I)$$

defined as $\overline{s \circ \Delta_C}(c + I) = \Delta_C(c) + (I \otimes C + C \otimes I)$ for $c \in C$. Let

$$\alpha : (C \otimes C)/(I \otimes C + C \otimes I) \to C/I \otimes C/I$$

be the isomorphism in the proof of Proposition 1.1.4, defined as

$$\alpha(a \otimes b + (I \otimes C + C \otimes I)) = (a + I) \otimes (b + I),$$

and let

$$\Delta_{C/I} : C/I \to C/I \otimes C/I$$

be the composition $\Delta_{C/I} = \alpha \circ \overline{s} \circ \Delta_C$ given as

$$\Delta_{C/I}(c + I) = \sum_{(c)} (c_{(1)} + I) \otimes (c_{(2)} + I).$$

Then $\Delta_{C/I}$ is K-linear and satisfies the coassociativity property since Δ_C does.

Moreover, $I \subseteq \ker(\epsilon_C)$ and so there is a map $\overline{\epsilon_C}/ : C/I \to K$, defined as $\overline{\epsilon_C}(c + I) = \epsilon_C(c)$. Put $\epsilon_{C/I} = \overline{\epsilon_C}$. Then $\epsilon_{C/I}$ is K-linear and satisfies the counit property since ϵ_C does. Thus $(C/I, \Delta_{C/I}, \epsilon_{C/I})$ is a K-coalgebra. $\qquad\square$

The coalgebra C/I is the **quotient coalgebra of C by I**.

Here is a construction of a quotient coalgebra. Let $K[x]$ be the coalgebra of Example 1.2.13. For $m \geq 1$, let I be the subspace of $K[x]$ generated by the basis

$$\{x^m - 1, x^{m+1} - 1, x^{m+2} - 1, \dots\}.$$

For $i \geq 0$,

$$\begin{aligned}
\Delta_{K[x]}(x^i - 1) &= x^i \otimes x^i - 1 \otimes 1 \\
&= (x^i - 1 + 1) \otimes (x^i - 1 + 1) - 1 \otimes 1 \\
&= (x^i - 1) \otimes (x^i - 1) + (x^i - 1) \otimes 1 + 1 \otimes (x^i - 1) + 1 \otimes 1 - 1 \otimes 1 \\
&= (x^i - 1) \otimes 1 + 1 \otimes (x^i - 1) + (x^i - 1) \otimes (x^i - 1).
\end{aligned}$$

Hence $\Delta_{K[x]}(I) \subseteq I + K[x] + K[x] \otimes I$. Also, $\epsilon_C(I) = 0$, and so I is a coideal of $K[x]$. The quotient coalgebra $K[x]/I$ is a vector space of dimension m on the basis $\{1, x, x^2, \dots, x^{m-1}\}$. It will become evident that $K[x]/I$ is isomorphic to the group ring Hopf algebra KC_m.

Let C be a K-coalgebra. A non-zero element c of C for which $\Delta_C(c) = c \otimes c$ is a **grouplike element** of C.

Proposition 1.2.16. *Let c be a grouplike element of the K-coalgebra C. Then* $\epsilon_C(c) = 1$.

Proof. If c is grouplike, then

$$\begin{aligned}
c &= s_1(\epsilon_C \otimes I_C)\Delta_C(c) \quad \text{by (1.6)} \\
&= s_1(\epsilon_C \otimes I_C)(c \otimes c) \\
&= \epsilon_C(c)c.
\end{aligned}$$

Thus $\epsilon_C(c)c = c = 1c$, so that $(\epsilon_C(c) - 1)c = 0$. Now since K is a field and $c \neq 0$, $\epsilon_C(c) - 1 = 0$, hence $\epsilon_C(c) = 1$. $\qquad \square$

In Example 1.2.12, the grouplike elements of V are precisely the basis $\mathcal{B}$, and in Example 1.2.13, the grouplike elements of $K[x]$ consist of $1, x, x^2, x^3 \ldots$.

Proposition 1.2.17. *Let $K[x]$ be the K-coalgebra as in Example 1.2.14. Then 1 is the only grouplike element in $K[x]$.*

Proof. Suppose $a_0 + a_1 x + a_2 x^2 + \cdots + a_n x^n$ is grouplike. Then $a_0 = 1$ by Proposition 1.2.16. Now

$$\Delta(1 + a_1 x + \cdots + a_n x^n) = 1 \otimes 1 + a_1(1 \otimes x + x \otimes 1)$$

$$+ a_2(1 \otimes x^2 + 2x \otimes x + x^2 \otimes 1)$$

$$+ \cdots + a_n \sum_{i=0}^{n} \binom{n}{i} x^i \otimes x^{n-i}$$

$$= (1 + a_1 x + \cdots + a_n x^n) \otimes (1 + a_1 x + \cdots + a_n x^n),$$

whence, $a_m = 0$ for $1 \leq m \leq n$. $\qquad \square$

Proposition 1.2.18. *Let C be a coalgebra and let $G(C)$ denote the set of grouplike elements of C. Then $G(C)$ is a linearly independent subset of C.*

Proof. If $G(C) = \emptyset$, then $G(C)$ is linearly independent. If $G(C)$ contains exactly one grouplike element, then this element is non-zero, and thus $G(C)$ is linearly independent. So we assume that $G(C)$ contains at least two elements.

Suppose that $G(C)$ is linearly dependent. Let $m \geq 1$ be the largest integer for which $S = \{g_1, g_2, \ldots, g_m\}$ is a linearly independent subset of $G(C)$. Then $G(C) \backslash S \neq \emptyset$, else $G(C)$ is linearly independent. Let $g \in G(C) \backslash S$. Then there exist scalars $r_i \in K$ with

$$g = r_1 g_1 + r_2 g_2 + \cdots + r_m g_m.$$

Since $g \neq 0$, $r_i \neq 0$ for at least one r_i, $1 \leq i \leq m$. Now,

$$\Delta_C(g) = g \otimes g = \sum_{i=1}^{m} \sum_{j=1}^{m} r_i r_j (g_i \otimes g_j)$$

At the same time,

$$\Delta_C(g) = \sum_{i=1}^{m} r_i (g_i \otimes g_i)$$

and so,

$$\sum_{i=1}^{m}\sum_{j=1}^{m} r_i r_j (g_i \otimes g_j) = \sum_{i=1}^{m} r_i (g_i \otimes g_i).$$

Since $\{g_i \otimes g_j\}_{1 \leq i,j \leq m}$ is a linearly independent subset of $C \otimes C$, $r_i r_j = 0$ for $i \neq j$, $1 \leq i, j \leq m$, and $r_i^2 = r_i$ for $1 \leq i \leq m$. Now, for any $r_i \neq 0$, we have $r_j = 0$ for $j \neq i$, thus $r_i \neq 0$ for exactly one i, and for this i, $r_i^2 = r_i$ implies $r_i = 1$. Thus $g = g_i$, which contradicts our choice of g. It follows that $G(C)$ is linearly independent. $\qquad\square$

The linear independence of grouplike elements will be used in §2.2 when we consider Myhill–Nerode bialgebras.

Let C, D be coalgebras. A K-linear map $\phi : C \to D$ is a **coalgebra homomorphism** if

$$(\phi \otimes \phi)\Delta_C(c) = \Delta_D(\phi(c))$$

and $\epsilon_C(c) = \epsilon_D(\phi(c))$ for all $c \in C$.

Proposition 1.2.19. *Let C be a K-coalgebra. Then the counit map $\epsilon_C : C \to K$ is a homomorphism of K-coalgebras.*

Proof. For $c \in C$,

$$(\epsilon_C \otimes \epsilon_C)\Delta_C(c) = (\epsilon_C \otimes I_K)(I_C \otimes \epsilon_C)\Delta_C(c)$$
$$= (\epsilon_C \otimes I_K)\left(\sum_{(c)} c_{(1)} \otimes \epsilon_C(c_{(2)}) \right)$$
$$= (\epsilon_C \otimes I_K)(c \otimes 1)$$
$$= \epsilon_C(c) \otimes 1$$
$$= \Delta_K(\epsilon_C(c)).$$

Moreover,

$$\epsilon_C(c) = \epsilon_K(\epsilon_C(c)),$$

thus ϵ_C is a coalgebra homomorphism. $\qquad\square$

A coalgebra homomorphism $\phi : C \to D$ that is injective and surjective is an **isomorphism of coalgebras**.

Proposition 1.2.20. *Let $\phi : C \to D$ be a homomorphism of K-coalgebras. If c is a grouplike element of C, then $\phi(c)$ is a grouplike element of D.*

Proof. Since ϕ is a coalgebra homomorphism,

$$\Delta_D(\phi(c)) = (\phi \otimes \phi)\Delta_C(c),$$

and since c is grouplike,

$$(\phi \otimes \phi)\Delta_C(c) = (\phi \otimes \phi)(c \otimes c) = \phi(c) \otimes \phi(c).$$

Thus $\phi(c)$ is grouplike. $\square$

Let $K[x]$ be the coalgebra as in Example 1.2.13, and let $K[x]'$ be the coalgebra as in Example 1.2.14. In view of Proposition 1.2.20, $K[x]$ and $K[x]'$ are not isomorphic as coalgebras since the only grouplike element of $K[x]'$ is 1 (Proposition 1.2.17).

The tensor product $C \otimes D$ of two coalgebras is again a coalgebra with comultiplication map

$$\Delta_{C \otimes D} : C \otimes D \to (C \otimes D) \otimes (C \otimes D)$$

defined by

$$
\begin{aligned}
\Delta_{C \otimes D}(c \otimes d) &= (I_C \otimes \tau \otimes I_D)(\Delta_C \otimes \Delta_D)(c \otimes d) \\
&= (I_C \otimes \tau \otimes I_D)(\Delta_C(c) \otimes \Delta_D(d)) \\
&= (I_C \otimes \tau \otimes I_D)\left(\sum_{(c),(d)} c_{(1)} \otimes c_{(2)} \otimes d_{(1)} \otimes d_{(2)} \right) \\
&= \sum_{(c),(d)} (c_{(1)} \otimes d_{(1)}) \otimes (c_{(2)} \otimes d_{(2)})
\end{aligned}
$$

for $c \in C$, $d \in D$. The counit map $\epsilon_{C \otimes D} : C \otimes D \to K$ is defined as

$$\epsilon_{C \otimes D}(c \otimes d) = \epsilon_C(c)\epsilon_D(d)$$

for $c \in C$, $d \in D$.

Example 1.2.21. Let C be the K-coalgebra of Example 1.2.10 and let D be the K-coalgebra of Example 1.2.11. Then $C \otimes_K D$ is a K-coalgebra on the basis $\{1 \otimes 1, 1 \otimes x, x \otimes 1, x \otimes x\}$. Comultiplication

$$\Delta_{C \otimes D} : C \otimes D \to (C \otimes D) \otimes (C \otimes D)$$

is defined by

$$\Delta_{C \otimes_K D}(1 \otimes 1) = (1 \otimes 1) \otimes (1 \otimes 1),$$

$$
\begin{aligned}
\Delta_{C \otimes_K D}(1 \otimes x) &= (I_C \otimes \tau \otimes I_D)(\Delta_C(1) \otimes \Delta_D(x)) \\
&= (I_C \otimes \tau \otimes I_D)((1 \otimes 1) \otimes (1 \otimes x + x \otimes 1)) \\
&= (I_C \otimes \tau \otimes I_D)(1 \otimes 1 \otimes 1 \otimes x + 1 \otimes 1 \otimes x \otimes 1) \\
&= (1 \otimes 1) \otimes (1 \otimes x) + (1 \otimes x) \otimes (1 \otimes 1),
\end{aligned}
$$

$$\Delta_{C\otimes_K D}(x \otimes 1) = (I_C \otimes \tau \otimes I_D)(\Delta_C(x) \otimes \Delta_D(1))$$
$$= (I_C \otimes \tau \otimes I_D)((x \otimes x) \otimes (1 \otimes 1))$$
$$= (x \otimes 1) \otimes (x \otimes 1),$$

$$\Delta_{C\otimes_K D}(x \otimes x) = (I_C \otimes \tau \otimes I_D)(\Delta_C(x) \otimes \Delta_D(x))$$
$$= (I_C \otimes \tau \otimes I_D)((x \otimes x) \otimes (1 \otimes x + x \otimes 1))$$
$$= (I_C \otimes \tau \otimes I_D)(x \otimes x \otimes 1 \otimes x + x \otimes x \otimes x \otimes 1)$$
$$= (x \otimes 1) \otimes (x \otimes x) + (x \otimes x) \otimes (x \otimes 1).$$

The counit map $\epsilon_{C\otimes_K D} : C \otimes_K D \to K$ is given as

$$\epsilon_{C\otimes_K D}(1 \otimes 1) = 1, \quad \epsilon_{C\otimes_K D}(1 \otimes x) = 0,$$

$$\epsilon_{C\otimes_K D}(x \otimes 1) = 1, \quad \epsilon_{C\otimes_K D}(x \otimes x) = 0.$$

1.3 Duality

In this section we consider the linear duals of algebras and coalgebras. We show that if $(C, \Delta_C, \epsilon_C)$ is a coalgebra, then $(C^*, m_{C^*}, \lambda_{C^*})$ is an algebra, where the maps m_{C^*} and λ_{C^*} are induced from the transpose of Δ_C and ϵ_C, respectively. We ask whether the converse of this statement is true (it isn't). However, if we replace A^* with a certain subspace A° called the finite dual, then A° is a coalgebra whenever A is an algebra. We show that the finite dual $K[x]^\circ$ can be identified with the collection of linearly recursive sequences of all orders over K.

$$* \quad * \quad *$$

Let C be a K-coalgebra and let C^* be its linear dual.

Proposition 1.3.1. *If C is a coalgebra, then C^* is an algebra.*

Proof. To show that C^* is a K-algebra we construct a multiplication map m_{C^*} and a unit map λ_{C^*} and show that they satisfy the associative and unit properties, respectively.

Recall that C is a triple $(C, \Delta_C, \epsilon_C)$ where $\Delta_C : C \to C \otimes C$ is K-linear and satisfies the coassociativity property (1.4), and $\epsilon_C : C \to K$ is K-linear and satisfies the counit property (1.5). The transpose of Δ_C is a K-linear map

$$\Delta_C^* : (C \otimes C)^* \to C^*$$

defined as

$$\Delta_C^*(\psi)(c) = \psi(\Delta_C(c)),$$

for $\psi \in (C \otimes C)^*, c \in C$.

By Corollary 1.1.8, $C^* \otimes C^* \subseteq (C \otimes C)^*$. Thus Δ_C^* restricts to a K-linear map $m_{C^*} : C^* \otimes C^* \to C^*$ defined as

$$\begin{aligned}
m_{C^*}(f \otimes g)(c) &= \Delta_C^*(f \otimes g)(c) \\
&= (f \otimes g)(\Delta_C(c)) \\
&= \sum_{(c)} f(c_{(1)})g(c_{(2)}).
\end{aligned}$$

for $f, g \in C^*, c \in C$. Let $I_{C^*} : C^* \to C^*$ be the identity map and let

$$I_{C^*} \otimes m_{C^*} : C^* \otimes C^* \otimes C^* \to C^* \otimes C^*,$$

be the map defined by

$$f \otimes g \otimes h \mapsto f \otimes m_{C^*}(g \otimes h),$$

and let

$$m_{C^*} \otimes I_{C^*} : C^* \otimes C^* \otimes C^* \to C^* \otimes C^*,$$

be the map defined by

$$f \otimes g \otimes h \mapsto m_{C^*}(f \otimes g) \otimes h,$$

for $f, g, h \in C^*$.

We are now ready to show that m_{C^*} satisfies the associative property. But this follows from the *coassociative property of* Δ_C! Indeed, for $f, g, h \in C^*, c \in C$,

$$\begin{aligned}
m_{C^*}(I_{C^*} \otimes m_{C^*})(f \otimes g \otimes h)(c) &= \Delta_C^*(I_{C^*} \otimes \Delta_C^*)(f \otimes g \otimes h)(c) \\
&= \Delta_C^*(f \otimes \Delta_C^*(g \otimes h))(c) \\
&= (f \otimes \Delta_C^*(g \otimes h))\Delta_C(c) \\
&= \sum_{(c)} f(c_{(1)})\Delta_C^*(g \otimes h)(c_{(2)}) \\
&= \sum_{(c)} f(c_{(1)})(g \otimes h)\Delta_C(c_{(2)}) \\
&= \sum_{(c)} f(c_{(1)}) \sum_{(c_{(2)})} g(c_{(2)(1)})h(c_{(2)(2)}).
\end{aligned}$$

Now,

$$\sum_{(c)} f(c_{(1)}) \sum_{(c_{(2)})} g(c_{(2)(1)}) h(c_{(2)(2)}) = \sum_{(c,c_{(2)})} f(c_{(1)}) g(c_{(2)(1)}) h(c_{(2)(2)})$$

$$= (f \otimes g \otimes h) \left(\sum_{(c,c_{(2)})} c_{(1)} \otimes c_{(2)(1)} \otimes c_{(2)(2)} \right)$$

$$= (f \otimes g \otimes h) \left(\sum_{(c,c_{(1)})} c_{(1)(1)} \otimes c_{(1)(2)} \otimes c_{(2)} \right)$$

$$= \sum_{(c,c_{(1)})} f(c_{(1)(1)}) g(c_{(1)(2)}) h(c_{(2)})$$

$$= \sum_{(c)} \sum_{(c_{(1)})} f(c_{(1)(1)}) g(c_{(1)(2)}) h(c_{(2)}).$$

Observe that the coassociativity of Δ_C is applied to get us from line 2 to line 3 above. We have established that

$$m_{C^*}(I_{C^*} \otimes m_{C^*})(f \otimes g \otimes h)(c) = \sum_{(c)} \sum_{(c_{(1)})} f(c_{(1)(1)}) g(c_{(1)(2)}) h(c_{(2)}).$$

To finish the calculation:

$$m_{C^*}(I_{C^*} \otimes m_{C^*})(f \otimes g \otimes h)(c) = \sum_{(c)} \sum_{(c_{(1)})} f(c_{(1)(1)}) g(c_{(1)(2)}) h(c_{(2)})$$

$$= \sum_{(c)} (f \otimes g) \Delta_C(c_{(1)}) h(c_{(2)})$$

$$= \sum_{(c)} \Delta_C^*(f \otimes g)(c_{(1)}) h(c_{(2)})$$

$$= (\Delta_C^*(f \otimes g) \otimes h) \Delta_C(c)$$

$$= \Delta_C^*(\Delta_C^*(f \otimes g) \otimes h)(c)$$

$$= \Delta_C^*(\Delta_C^* \otimes I_{C^*})(f \otimes g \otimes h)(c)$$

$$= m_{C^*}(m_{C^*} \otimes I_{C^*})(f \otimes g \otimes h)(c).$$

Thus m_{C^*} satisfies the associative property.

The transpose of the counit map of C is

$$\epsilon_C^* : K^* \to C^*$$

defined as $\epsilon_C^*(f)(c) = f(\epsilon_C(c))$ for $f \in K^*$, $c \in C$. Identifying $K = K^*$, we have $\epsilon_C^* : K \to C^*$ defined as

$$\epsilon_C^*(r)(c) = r(\epsilon_C(c)) = r\epsilon_C(c),$$

for $r \in K$, $c \in C$. Set $\lambda_{C^*} = \epsilon_C^*$ and define maps $I_{C^*} \otimes \lambda_{C^*} : C^* \otimes K \to C^* \otimes C^*$, $f \otimes r \mapsto f \otimes \lambda_{C^*}(r)$, $\lambda_{C^*} \otimes I_{C^*} : K \otimes C^* \to C^* \otimes C^*$, $r \otimes f \mapsto \lambda_{C^*}(r) \otimes f$, for $f \in C^*$, $r \in K$.

We now show that the counit property of ϵ_C implies the unit property for λ_{C^*}. To this end, for $f \in C^*$, $r \in K$, $c \in C$,

$$
\begin{aligned}
m_{C^*}(I_{C^*} \otimes \lambda_{C^*})(f \otimes r)(c) &= \Delta_C^*(I_{C^*} \otimes \epsilon_C^*)(f \otimes r)(c) \\
&= \Delta_C^*(f \otimes \epsilon_C^*(r))(c) \\
&= (f \otimes \epsilon_C^*(r))(\Delta_C(c)) \\
&= \sum_{(c)} f(c_{(1)})\epsilon_C^*(r)(c_{(2)}) \\
&= \sum_{(c)} f(c_{(1)})r(\epsilon_C(c_{(2)})).
\end{aligned}
$$

Thus,

$$m_{C^*}(I_{C^*} \otimes \lambda_{C^*})(f \otimes r)(c) = r\sum_{(c)} f(c_{(1)})\epsilon_C(c_{(2)}).$$

Proceeding with the calculation:

$$
\begin{aligned}
m_{C^*}(I_{C^*} \otimes \lambda_{C^*})(f \otimes r)(c) &= r\sum_{(c)} f(c_{(1)})\epsilon_C(c_{(2)}) \\
&= r\sum_{(c)} \epsilon_C(c_{(2)})f(c_{(1)}) \\
&= r\sum_{(c)} f(\epsilon_C(c_{(2)})c_{(1)}) \\
&= rf\left(\sum_{(c)} \epsilon_C(c_{(2)})c_{(1)} \right) \\
&= rf(c).
\end{aligned}
$$

Note that the counit property of ϵ_C is applied to get us from line 4 to line 5 above. In a similar manner, we obtain

$$m_{C^*}(\lambda_{C^*} \otimes I_{C^*})(r \otimes f) = rf.$$

Thus λ_{C^*} satisfies the unit property and we have established that $(C^*, m_{C^*}, \lambda_{C^*})$ is an algebra. We have $\lambda_{C^*}(1_K)(c) = \epsilon_C(c), \forall c$, and so, ϵ_C is the unique element of C^* for which $\epsilon_C f = f = f \epsilon_C$ for all $f \in C^*$. □

As we have just seen, if $(C, \Delta_C, \epsilon_C)$ is a coalgebra, then $(C^*, m_{C^*}, \lambda_{C^*})$ is an algebra. Now suppose (A, m_A, λ_A) is a K-algebra. Then one may wonder if A^* is a K-coalgebra. The transpose of the multiplication map $m_A : A \otimes_K A \to A$ is $m_A^* : A^* \to (A \otimes_K A)^*$. However, if A is infinite dimensional over K, then $A^* \otimes_K A^*$ is a proper subset of $(A \otimes_K A)^*$. Hence, in the case that A is infinite dimensional, m_A^* may not induce the required comultiplication map $A^* \to A^* \otimes_K A^*$. Indeed, here is an example where $m_A^*(A^*) \nsubseteq A^* \otimes_K A^*$.

Example 1.3.2. Let $A = K[x]$ be the K-algebra of polynomials in x. Let $m_{K[x]} : K[x] \otimes_K K[x] \to K[x]$ be the multiplication map with transpose $m_{K[x]}^* : K[x]^* \to (K[x] \otimes_K K[x])^*$. Let $\{s_n\}$ be the sequence in K given as

$$1, 0, 0, 1, 1, 1, 0, 0, 0, 0, \ldots$$

(that is, one 1, followed by two 0's, followed by three 1's, and so on). Let f be the element in $K[x]^*$ defined as

$$f = 1e_0 + 0e_1 + 0e_2 + 1e_3 + 1e_4 + 1e_5 + 0e_6 + 0e_7 + 0e_8 + 0e_9$$
$$+ 1e_{10} + 1e_{11} + 1e_{12} + 1e_{13} + 1e_{14} + \cdots$$

where $e_i(x^j) = \delta_{i,j}, i, j \geq 0$. Then $m_{K[x]}^*(f) \nsubseteq K[x]^* \otimes_K K[x]^*$.

In order for the transpose m_A^* to serve as a comultiplication map we must replace A^* with a certain subspace of A^*, called the finite dual A° of A^*.

Let (A, m_a, λ_A) be a K-algebra. Then A is a K-vector space and a ring. An ideal I of A has **finite codimension** (or is **cofinite**) if the quotient space A/I is finite dimensional. Let f be an element of A^* and let S be a subset of A. Then **vanishes on** S if $f(s) = 0$ for all $s \in S$.

Definition 1.3.3. Let A be a K-algebra. The **finite dual** A° **of** A is the subspace of A^* defined as

$$A^\circ = \{f \in A^* : f \text{ vanishes on some ideal } I \subseteq A \text{ of finite codimension}\}$$

Example 1.3.4. Let $e_i \in K[x]^*$ be defined as $e_i(x^j) = \delta_{i,j}$ for $i, j \geq 0$. Then $e_i \in K[x]^\circ$ since e_i vanishes on the ideal (x^{i+1}) and $\dim(K[x]/(x^{i+1})) = i + 1$.

Proposition 1.3.5. *If A is finite dimensional as a K-vector space, then $A^\circ = A^*$.*

Proof. Exercise. □

Our goal is to show that if A is an algebra, then A° is a coalgebra (Proposition 1.3.9). We need three propositions.

Proposition 1.3.6. *Let I be an ideal of A and let $s : A \to A/I$ be the canonical surjection of vector spaces. Let $s^* : (A/I)^* \to A^*$ be the transpose defined as $s^*(f)(a) = f(s(a))$, for all $f \in (A/I)^*$, $a \in A$. Then s^* is an injection.*

Proof. Let $f, g \in (A/I)^*$. Then for $a \in A$, $s^*(f)(a) = s^*(g)(a)$ implies that $f(s(a)) = g(s(a))$. Thus $f(a + I) = g(a + I)$, and so $f = g$. $\qquad\square$

Proposition 1.3.7. *Let $f \in A^\circ$ and suppose that f vanishes on the ideal I of A. Then there exists a unique element $\bar{f} \in (A/I)^*$ for which $s^*(\bar{f}) = f$.*

Proof. The element $f \in A^\circ \subseteq A^*$ is a K-module homomorphism $f : A \to K$ and since $f(I) = 0$, $I \subseteq \ker(f)$. Let $s : A \to A/I$ denote the canonical surjection. Then by the universal mapping property for kernels there exists a K-module homomorphism $\bar{f} : A/I \to K$ for which $\bar{f}(s(a)) = f(a)$ for all $a \in A$. Thus $s^*(\bar{f})(a) = f(a)$ for all $a \in A$. $\qquad\square$

Let A be a K-algebra with multiplication map $m_A : A \otimes A \to A$. Let $m_A^* : A^* \to (A \otimes A)^*$ be the transpose map defined as

$$m_A^*(f)(a \otimes b) = f(m_A(a \otimes b)) = f(ab).$$

Let $A^\circ \subseteq A^*$ be the finite dual of A.

Proposition 1.3.8. $m_A^*(A^\circ) \subseteq A^\circ \otimes A^\circ$.

Proof. Let $f \in A^\circ$. Then f vanishes on some ideal $I \subseteq A$ with $\dim(A/I) < \infty$. By Proposition 1.3.7 there exists a unique element $\bar{f} \in (A/I)^*$ for which $s^*(\bar{f}) = f$. Let

$$m_{A/I} : A/I \otimes A/I \to A/I,$$

$(a + I) \otimes (b + I) \mapsto ab + I$, be the multiplication map of the K-algebra A/I, cf. Proposition 1.2.7. The transpose of $m_{A/I}$ is

$$m_{A/I}^* : (A/I)^* \to (A/I \otimes A/I)^*,$$

which, since A/I is finite dimensional, becomes

$$m_{A/I}^* : (A/I)^* \to (A/I)^* \otimes (A/I)^*.$$

Now,

$$
\begin{aligned}
m_A^*(f)(a \otimes b) &= m_A^*(s^*(\bar{f}))(a \otimes b) \\
&= s^*(\bar{f})(m_A(a \otimes b)) \\
&= s^*(\bar{f})(ab) \\
&= \bar{f}(s(ab)) \\
&= \bar{f}(ab + I) \\
&= \bar{f}(m_{A/I}((a + I) \otimes (b + I))).
\end{aligned}
$$

Continuing with this calculation, we obtain

$$m_A^*(f)(a \otimes b) = \bar{f}(m_{A/I}((a+I) \otimes (b+I)))$$
$$= m_{A/I}^*(\bar{f})((a+I) \otimes (b+I))$$
$$= m_{A/I}^*(\bar{f})(s(a) \otimes s(b)).$$

Note that $m_{A/I}^*(\bar{f}) = \sum_{i=1}^m f_i \otimes g_i$ for elements $f_i, g_i \in (A/I)^*$. Thus,

$$m_A^*(f)(a \otimes b) = m_{A/I}^*(\bar{f})(s(a) \otimes s(b))$$

$$= \left(\sum_{i=1}^m f_i \otimes g_i \right)(s(a) \otimes s(b))$$

$$= \sum_{i=1}^m f_i(s(a)) \otimes g_i(s(b))$$

$$= \sum_{i=1}^m s^*(f_i)(a) \otimes s^*(g_i)(b)$$

$$= \left(\sum_{i=1}^m s^*(f_i) \otimes s^*(g_i) \right)(a \otimes b),$$

Thus

$$m_A^*(f) = \sum_{i=1}^m s^*(f_i) \otimes s^*(g_i).$$

It remains to show that $s^*(f_i), s^*(g_i) \in A^\circ$. To this end, let $q \in I$, then $s^*(f_i)(q) = f_i(s(q)) = f_i(I) = 0 \in K$, and so $s^*(f_i)$ vanishes on I, hence $s^*(f_i) \in A^\circ$. A similar argument shows that $s^*(g_i) \in A^\circ$. Consequently, $m_A^*(f) \subseteq A^\circ \otimes A^\circ$.   □

The transpose of the map

$$I_A \otimes m_A : A \otimes A \otimes A \to A \otimes A$$

restricted to $A^* \otimes A^*$ is the map

$$I_A^* \otimes m_A^* = I_{A^*} \otimes m_A^* : A^* \otimes A^* \to (A \otimes A \otimes A)^*.$$

Likewise, the transpose of $m_A \otimes I_A$ restricted to $A^* \otimes A^*$ is

$$m_A^* \otimes I_A^* = m_A^* \otimes I_{A^*} : A^* \otimes A^* \to (A \otimes A \otimes A)^*.$$

Proposition 1.3.9. *If A is an algebra, then A° is a coalgebra.*

Proof. Let $m_A^* : A^* \to (A \otimes A)^*$ be the transpose of the multiplication map m_A. By Proposition 1.3.8, $m_A^*(A^\circ) \subseteq A^\circ \otimes A^\circ$. Let Δ_{A° denote the restriction of m_A^* to A°. Then $\Delta_{A^\circ} : A^\circ \to A^\circ \otimes A^\circ$ is a K-linear map defined as $\Delta_{A^\circ}(f) = m_A^*(f)$ for $f \in A^\circ$. The first step is to show that the associative property of m_A implies the coassociative condition for Δ_{A°. For $f \in A^\circ$, $a, b, c \in A$, we have

$$
\begin{aligned}
(I_{A^\circ} \otimes \Delta_{A^\circ})\Delta_{A^\circ}(f)(a \otimes b \otimes c) &= (I_A^* \otimes m_A^*)m_A^*(f)(a \otimes b \otimes c) \\
&= m_A^*(f)(I_A \otimes m_A)(a \otimes b \otimes c) \\
&= m_A^*(f)(a \otimes bc) \\
&= f(m_A(a \otimes bc)) \\
&= f(a(bc)) \\
&= f((ab)c).
\end{aligned}
$$

We have applied the associative property of m_A to move from line 5 to line 6 above. Now,

$$
\begin{aligned}
(I_{A^\circ} \otimes \Delta_{A^\circ})\Delta_{A^\circ}(f)(a \otimes b \otimes c) &= f((ab)c) \\
&= f(m_A(ab \otimes c)) \\
&= m_A^*(f)(ab \otimes c) \\
&= m_A^*(f)(m_A \otimes I_A)(a \otimes b \otimes c) \\
&= (m_A^* \otimes I_A^*)m_A^*(f)(a \otimes b \otimes c) \\
&= (\Delta_{A^\circ} \otimes I_{A^\circ})\Delta_{A^\circ}(f)(a \otimes b \otimes c),
\end{aligned}
$$

and so, the cooassociative property holds for Δ_{A°.

For the counit map of A°, we consider the dual map $\lambda_A^* : A^* \to K^* = K$. We let $\epsilon_{A^\circ} : A^\circ \to K$ denote the restriction of λ_A^* to A°. For $f \in A^\circ$, $r \in K$,

$$
\epsilon_{A^\circ}(f)(r) = f(\lambda_A(r)) = f(r1_A) = rf(1_A) = f(1_A)(r)
$$

and so, $\epsilon_{A^\circ}(f) = f(1_A)$. We show that the unit property of λ_A implies that ϵ_{A° satisfies the counit property. For $f \in A^\circ$, $r \in K$, $a \in A$,

$$
\begin{aligned}
(\epsilon_{A^\circ} \otimes I_{A^\circ})\Delta_{A^\circ}(f)(r \otimes a) &= (\lambda_A^* \otimes I_A^*)m_A^*(f)(r \otimes a) \\
&= m_A^*(f)(\lambda_A \otimes I_A)(r \otimes a) \\
&= f(m_A(\lambda_A \otimes I_A)(r \otimes a)) \\
&= f(ra)
\end{aligned}
$$

$$= rf(a)$$
$$= (1 \otimes f)(r \otimes a).$$

The reader should note that we have applied the unit property of λ_A to move from line 3 to line 4 above. Thus

$$(\epsilon_{A^\circ} \otimes I_{A^\circ})\Delta_{A^\circ}(f) = 1 \otimes f.$$

In a similar manner, one obtains

$$(I_{A^\circ} \otimes \epsilon_{A^\circ})\Delta_{A^\circ}(f) = f \otimes 1.$$

Thus ϵ_{A° satisfies the counit condition and it follows that $(A^\circ, \Delta_{A^\circ}, \epsilon_{A^0})$ is a coalgebra. $\qquad\square$

If A is a K-algebra that is finite dimensional as a K vector space, then $A^\circ = A^*$ and A^* is a K-coalgebra. For $a, b \in A, f \in A^*$, we have the formula

$$\begin{aligned}
f(ab) &= f(m_A(a \otimes b)) \\
&= m_A^*(f)(a \otimes b) \\
&= \left(\sum_{(f)} f_{(1)} \otimes f_{(2)} \right)(a \otimes b) \\
&= \sum_{(f)} f_{(1)}(a)f_{(2)}(b).
\end{aligned} \qquad (1.9)$$

Here is an application of formula (1.9). Let S be a finite monoid, $|S| = m$, and let KS be the monoid ring. The K-algebra KS is finite dimensional over K on the basis $\{1 = g_0, g_1, g_2, \ldots, g_{m-1}\}$. Now $KS^\circ = KS^*$ with K-basis $\{e_1, e_{g_1}, \ldots, e_{g_{m-1}}\}$, $e_{g_i}(g_j) = \delta_{i,j}$.

Proposition 1.3.10. *Let* $(KS^*, \Delta_{KS^*}, \epsilon_{KS^*})$ *be the coalgebra as in the preceding text. Then*

$$\Delta_{KS^*}(e_{g_i}) = \sum_{g_j g_k = g_i} e_{g_j} \otimes e_{g_k},$$

for $i = 0, 1, \ldots, m - 1$.

Proof. Note that $\{e_{g_j} \otimes e_{g_k}\}, 0 \le j, k \le m - 1$, is a K-basis for $KS^* \otimes KS^*$. Fix i, $0 \le i \le m - 1$. There exist elements $a_{i,j',k'} \in K$ for which

$$\Delta_{KS^*}(e_{g_i}) = \sum_{j',k'=0}^{m-1} a_{i,j',k'}(e_{g_{j'}} \otimes e_{g_{k'}}).$$

Let j, k be so that $g_i = g_j g_k$. Then by (1.9),

$$1 = e_{g_i}(g_i)$$

$$= e_{g_i}(g_j g_k)$$

$$= \sum_{j',k'=0}^{m-1} a_{i,j',k'} e_{g_{j'}}(g_j) e_{g_{k'}}(g_k)$$

$$= a_{i,j,k}.$$

If i, j are so that $g_i \neq g_j g_k$, then

$$0 = e_{g_i}(g_j g_k)$$

$$= \sum_{j',k'=0}^{m-1} a_{i,j',k'} e_{g_{j'}}(g_j) e_{g_{k'}}(g_k)$$

$$= a_{i,j,k}.$$

It follows that

$$\Delta_{KS^*}(e_{g_i}) = \sum_{g_j g_k = g_i} e_{g_j} \otimes e_{g_k},$$

for $i = 0, 1, \ldots, m - 1$. $\qquad\qquad\qquad\qquad\qquad\qquad\qquad\qquad\qquad\qquad\qquad\square$

Also:

Proposition 1.3.11. *If (A, m_A, λ_A) be a commutative algebra, then $(A^\circ, \Delta_{A^\circ}, \epsilon_{A^\circ})$ is a cocommutative coalgebra. If $(C, \Delta_C, \epsilon_C)$ is a cocommutative coalgebra, then $(C^*, m_{C^*}, \lambda_{C^*})$ is a commutative algebra.*

Proof. Let $\tau : A \otimes A \to A \otimes A$, $a \otimes b \mapsto b \otimes a$, be the twist map. The transpose τ^* restricted to $A^* \otimes A^*$ is the twist map on $A^* \otimes A^*$. Now, to prove the first statement, let $f \in A^\circ$, $a, b \in A$. Then

$$\Delta_{A^\circ}(f)(a \otimes b) = f(m_A(a \otimes b))$$

$$= f(m_A(\tau(a \otimes b)))$$

$$= \tau^*(\Delta_{A^\circ}(f))(a \otimes b)$$

$$= \tau(\Delta_{A^\circ}(f))(a \otimes b),$$

and so, A° is cocommutative.

The second statement is left as an exercise. $\qquad\qquad\qquad\qquad\qquad\qquad\square$

Let $K[x]$ denote the algebra of polynomials over the field K. As in Example 1.3.2, the collection of sequences $\{s_n\}_{n=0}^\infty$ over K can be identified with the linear dual $K[x]^*$. Indeed, the sequence $\{s_n\}_{n=0}^\infty$ corresponds to the infinite sum $s = \sum_{n=0}^\infty s_n e_n$

where $e_i(x^j) = \delta_{i,j}$, $\forall i, j$. The next proposition ties the development of the finite dual to linearly recursive sequences in a rather elegant way.

Proposition 1.3.12. *Let K be a field. The collection of kth-order linearly recursive sequences over K of all orders $k > 0$ can be identified with the finite dual $K[x]^\circ$.*

Proof. Let $\{s_n\}$ be a kth-order linearly recursive sequence over K of order $k > 0$ with recurrence relation

$$s_{n+k} = a_{k-1}s_{n+k-1} + a_{k-2}s_{n+k-2} + \cdots + a_1 s_{n+1} + a_0 s_n, \quad n \geq 0, \qquad (1.10)$$

and characteristic polynomial

$$f(x) = x^k - a_{k-1}x^{k-1} - a_{k-2}x^{k-2} - \cdots - a_1 x - a_0, \qquad (1.11)$$

for $a_0, a_1, a_2, \ldots, a_{k-1} \in K$. We identify $\{s_n\}$ with the element $s = \sum_{n=0}^{\infty} s_n e_n \in K[x]^*$. Now,

$$s(f(x)) = \left(\sum_{n=0}^{\infty} s_n e_n \right)(x^k - a_{k-1}x^{k-1} - a_{k-2}x^{k-2} - \cdots - a_1 x - a_0)$$

$$= s_k - a_{k-1}s_{k-1} - a_{k-2}s_{k-2} - \cdots - a_1 s_1 - a_0 s_0$$

$$= 0.$$

It follows that $s(g(x)f(x)) = 0$ for all $g(x) \in K[x]$, and so s vanishes on the principal ideal $I = (f(x))$ of $K[x]$. One has $\dim(K[x]/I) = k$ and so $s \in K[x]^\circ$.

On the other hand, let $s = \sum_{n=0}^{\infty} s_n e_n \in K[x]^\circ$. Then s vanishes on an ideal $I \subseteq K[x]$ of finite codimension. Since $K[x]$ is a PID, $I = (f(x))$ for some monic polynomial over K of degree k. Then as one can easily check, $\{s_n\}$ is a kth-order linearly recursive sequence over K with characteristic polynomial $f(x)$. $\square$

By Proposition 1.3.9, $K[x]^\circ$, the linearly recursive sequences over K of all orders, is a coalgebra $(K[x]^\circ, \Delta_{K[x]^\circ}, \epsilon_{K[x]^\circ})$.

We ask: given $s = \{s_n\} \in K[x]^\circ$, how does one compute $\Delta_{K[x]^\circ}(s) \in K[x]^\circ \otimes K[x]^\circ$? Suppose s vanishes on the ideal $I = (f(x))$ of $K[x]$ where $f(x)$ is the characteristic polynomial of s. Let $c : K[x] \to K[x]/(f(x))$ denote the canonical surjection. In view of the method used in the proof of Proposition 1.3.8, we first find $\bar{s} \in (K[x]/(f(x)))^*$ so that $c^*(\bar{s}) = s$. We next compute

$$m^*_{K[x]/(f(x))}(\bar{s}) = \sum_{i=1}^{m} f_i \otimes g_i,$$

where $f_i, g_i \in (K[x]/(f(x)))^*$. Finally, we have

$$\Delta_{K[x]^\circ}(s) = \sum_{i=1}^{m} c^*(f_i) \otimes c^*(g_i) \in K[x]^\circ \otimes K[x]^\circ.$$

To see how this works in a modest example, take $K = \mathrm{GF}(2)$. Let $s = \{s_n\}$ be the 2nd-order linearly recursive sequence in K with characteristic polynomial $f(x) = x^2 + x + 1$ and initial state vector $s_0 = 11$ (a Fibonacci sequence). Let α be a zero of $f(x)$. Then $\{1, \alpha\}$ is a K-basis for $K[x]/(f(x)) = \mathrm{GF}(4)$ and $\{\varepsilon_1, \varepsilon_\alpha\}$ is a K-basis for $(K[x]/(f(x)))^* = \mathrm{GF}(4)^*$ with $\varepsilon_{\alpha^i}(\alpha^j) = \delta_{i,j}$, $0 \le i, j \le 1$.

Now, with $\{e_i\}$ so that $e_i(x^j) = \delta_{i,j}$ for $i, j \ge 0$, the sequence $s = \{s_n\}$ can be written as an element of $K[x]^\circ$:

$$s = 1 \cdot e_0 + 1 \cdot e_1 + 0 \cdot e_2 + 1 \cdot e_3 + 1 \cdot e_4 + \cdots$$

Let $\bar{s} = 1 \cdot \varepsilon_1 + 1 \cdot \varepsilon_\alpha$. We claim that $c^*(\bar{s}) = s$. To this end observe that for all $j \ge 0$,

$$
\begin{aligned}
c^*(\bar{s})(x^j) &= \bar{s}(c(x^j)) \\
&= \bar{s}(x^j + (f(x))) \\
&= \bar{s}(\alpha^j) \\
&= s_j,
\end{aligned}
$$

and so, $c^*(\bar{s}) = s$.

Our next step is to compute $m^*_{\mathrm{GF}(4)}(\bar{s})$. But note that

$$m^*_{\mathrm{GF}(4)}(\bar{s}) = m^*_{\mathrm{GF}(4)}(\varepsilon_1) + m^*_{\mathrm{GF}(4)}(\varepsilon_\alpha),$$

and so this depends on the computation of $m^*_{\mathrm{GF}(4)}(\varepsilon_1)$ and $m^*_{\mathrm{GF}(4)}(\varepsilon_\alpha)$

To compute $m^*_{\mathrm{GF}(4)}(\varepsilon_1)$ and $m^*_{\mathrm{GF}(4)}(\varepsilon_\alpha)$ we use the formula (1.9) and the idea of Proposition 1.3.10. First note that

$$
\begin{aligned}
m^*_{\mathrm{GF}(4)}(\varepsilon_1) &= c_{0,0,0}(\varepsilon_1 \otimes \varepsilon_1) + c_{0,0,1}(\varepsilon_1 \otimes \varepsilon_\alpha) + c_{0,1,0}(\varepsilon_\alpha \otimes \varepsilon_1) \\
&\quad + c_{0,1,1}(\varepsilon_\alpha \otimes \varepsilon_\alpha).
\end{aligned}
$$

for some bits $c_{0,i,j}$, $0 \le i, j \le 1$. Thus,

$$
\begin{aligned}
\varepsilon_1(ab) &= c_{0,0,0}\varepsilon_1(a)\varepsilon_1(b) + c_{0,0,1}\varepsilon_1(a)\varepsilon_\alpha(b) + c_{0,1,0}\varepsilon_\alpha(a)\varepsilon_1(b) \\
&\quad + c_{0,1,1}\varepsilon_\alpha(a)\varepsilon_\alpha(b),
\end{aligned}
$$

for all $a, b \in \mathrm{GF}(4)$. Now since

$$1 = 1 \cdot 1 = \alpha(1 + \alpha) = (1 + \alpha)\alpha,$$

and

$$\alpha = 1 \cdot \alpha = \alpha \cdot 1 = (1 + \alpha)(1 + \alpha),$$

in $\mathrm{GF}(4)$, we conclude that

$$m^*_{\mathrm{GF}(4)}(\varepsilon_1) = \varepsilon_1 \otimes \varepsilon_1 + \varepsilon_\alpha \otimes \varepsilon_\alpha.$$

By a similar method we also obtain

$$m^*_{GF(4)}(\varepsilon_\alpha) = \varepsilon_1 \otimes \varepsilon_\alpha + \varepsilon_\alpha \otimes \varepsilon_1 + \varepsilon_\alpha \otimes \varepsilon_\alpha.$$

Thus

$$m^*_{GF(4)}(\overline{s}) = \varepsilon_1 \otimes \varepsilon_1 + \varepsilon_1 \otimes \varepsilon_\alpha + \varepsilon_\alpha \otimes \varepsilon_1.$$

Finally,

$$\Delta_{K[x]^\circ}(s) = c^*(\varepsilon_1) \otimes c^*(\varepsilon_1) + c^*(\varepsilon_1) \otimes c^*(\varepsilon_\alpha) + c^*(\varepsilon_\alpha) \otimes c^*(\varepsilon_1)$$
$$= r \otimes r + r \otimes t + t \otimes r,$$

where $r = \{r_n\}$ is the Fibonacci sequence in GF(2) with initial state vector $\mathbf{r_0} = 10$, and $t = \{t_n\}$ is the Fibonacci sequence in GF(2) with initial state vector $\mathbf{t_0} = 01$.

1.4 Chapter Exercises

Exercises for §1.1

1. Let $r \geq 1$ and let $M_1, M_2, \ldots, M_r, M_{r+1}$ be R-modules. Show that

$$M_1 \otimes M_2 \otimes \cdots \otimes M_r \otimes M_{r+1} \cong (M_1 \otimes M_2 \otimes \cdots \otimes M_r) \otimes M_{r+1}$$

2. Let $r, s \geq 1$ and let $M_1, M_2, \ldots, M_{r+s}$ be R-modules. Use induction on s to prove that

$$M_1 \otimes M_2 \otimes \cdots \otimes M_{r+s} \cong (M_1 \otimes M_2 \otimes \cdots \otimes M_r) \otimes (M_{r+1} \otimes M_{r+2} \otimes \cdots \otimes M_{r+s}).$$

(The trivial case $s = 1$ is Exercise 1.)

3. Let $f_1 : V_1 \to V_1'$ and $f_2 : V_2 \to V_2'$ be linear transformations of vector spaces and let

$$(f_1 \otimes f_2) : V_1 \otimes V_2 \to V_1' \otimes V_2'$$

be the map defined as $a \otimes b \mapsto f_1(a) \otimes f_2(b)$. Prove that the transpose

$$(f_1 \otimes f_2)^* : (V_1' \otimes V_2')^* \to (V_1 \otimes V_2)^*$$

restricted to $(V_1')^* \otimes (V_2')^*$ is the map $f_1^* \otimes f_2^*$.

Exercises for §1.2

4. Let (A, m_A, λ_A) be a K-algebra, let I be an ideal of A, and let $(A/I, m_{A/I}, \lambda_{A/I})$ be the quotient algebra. Verify that the maps $m_{A/I}$ and $\lambda_{A/I}$ satisfy the associative and unit properties, respectively.

5. Let K be a field, let C be a K-coalgebra, and let $c \in C$. Prove that

$$(I_C \otimes s_1)(I_C \otimes \epsilon_C \otimes I_C)\left(\sum_{(c)} c_{(1)} \otimes c_{(2)} \otimes c_{(3)} \right)$$

$$= (s_1 \otimes I_C)(\epsilon_C \otimes I_C \otimes I_C)\left(\sum_{(c)} c_{(1)} \otimes c_{(2)} \otimes c_{(3)} \right).$$

6. Let $(K[x], \Delta_{K[x]}, \epsilon_{K[x]})$ be the coalgebra of Example 1.2.14. Verify that the maps $\Delta_{K[x]}$ and $\epsilon_{K[x]}$ satisfy the coassociative and counit properties, respectively.
7. Let $(C, \Delta_C, \epsilon_C)$ be a coalgebra with $\epsilon_C(C) = K$. Let $I = \ker(\epsilon_C)$. Prove that I is a coideal.
8. Let I and J be coideals of the coalgebra C. Show that

$$I + J = \{a + b : a \in I, b \in J\}$$

is a coideal of C.
9. Let $(C, \Delta_C, \epsilon_C)$ be a K-coalgebra, let I be a coideal of C and let $(C/I, \Delta_{C/I}, \epsilon_{C/I})$ be the quotient coalgebra. Verify that the maps $\Delta_{C/I}$ and $\epsilon_{C/I}$ satisfy the coassociative and counit properties, respectively.
10. Let C be a K-coalgebra. Define a coalgebra structure on $C \otimes C \otimes C$ in two different ways. Are the resulting K-coalgebras isomorphic as coalgebras?

Exercises for §1.3

11. Prove Proposition 1.3.5.
12. Prove the second statement of Proposition 1.3.11.
13. Let $K = GF(2)$ and let $\{s_n\}$ be a 3rd-order linearly recursive sequence in $GF(2)$ with recurrence relation

$$s_{n+3} = s_{n+1} + s_n,$$

characteristic polynomial $f(x) = x^3 + x + 1$ and initial state vector $s_0 = 111$. Compute $\Delta_{K[x]^\circ}(s) \in K[x]^\circ \otimes K[x]^\circ$.
14. Let $K = GF(2)$ and let $\{s_n\}$ be a 5th-order linearly recursive sequence in $GF(2)$ with recurrence relation

$$s_{n+5} = s_n,$$

characteristic polynomial $f(x) = x^5 + 1$ and initial state vector $s_0 = 10001$. Compute $\Delta_{K[x]^\circ}(s) \in K[x]^\circ \otimes K[x]^\circ$.

Chapter 2
Bialgebras

In this chapter we consider bialgebras—vector spaces that are both algebras and coalgebras. We give some basic examples and show that if B is a bialgebra, then B° is a bialgebra. We show that $K[x]$ is a bialgebra in exactly two distinct ways, and so $K[x]^\circ$ is a bialgebra in two distinct ways. Consequently, we can multiply linearly recursive sequences in two different ways, namely, the Hadamard product and the Hurwitz product.

Next we give an application of bialgebras to theoretical computer science. We introduce finite automata, and prove the Myhill–Nerode theorem which tells us precisely when a language is accepted by a finite automaton. We then generalize the Myhill–Nerode theorem to an algebraic setting in which a certain finite dimensional bialgebra (a Myhill–Nerode bialgebra) plays the role of the finite automaton that accepts the language. We see that a Myhill–Nerode bialgebra determines a finite automaton and a finite automaton determines a Myhill–Nerode bialgebra. We can think of finite automata and languages in terms of the algebraic properties of their Myhill–Nerode bialgebras. For instance, if two languages determine isomorphic Myhill–Nerode bialgebras, then these languages are related in some manner.

In the final section of the chapter, we introduce regular sequences; these are sequences that generalize linearly recursive sequences over a Galois field.

2.1 Introduction to Bialgebras

In this section we introduce bialgebras and define biideals, quotient bialgebras, and bialgebra homomorphisms. We show how a bialgebra B can act on an algebra A giving A the structure of a left B-module algebra, and how the bialgebra can act on a coalgebra C so that C is a right B-module coalgebra. We define a certain right action of B on the algebra B^*, the right translate $f \leftharpoonup a$ of $f \in B^*$ by $a \in B$. (In fact, the right translate action endows B^* with the structure of a right B-module

© Springer International Publishing Switzerland 2015
R.G. Underwood, *Fundamentals of Hopf Algebras*, Universitext,
DOI 10.1007/978-3-319-18991-8_2

algebra.) We state an important result which says that $\dim(f \leftharpoonup B) < \infty$ if and only if $f \in B^\circ$. As a consequence we show that if B is a bialgebra, then the finite dual B° is a bialgebra. We show that $K[x]$ is a bialgebra in exactly two distinct ways, thus $K[x]^\circ$ has two distinct structures as a bialgebra. The resulting multiplications on $K[x]^\circ$ are the Hadamard product and the Hurwitz product of linearly recursive sequences.

* * *

Definition 2.1.1. A **K-bialgebra** is a K-vector space B together with maps m_B, λ_B, Δ_B, ϵ_B that satisfy the following conditions:

(i) (B, m_B, λ_B) is a K-algebra and $(B, \Delta_B, \epsilon_B)$ is a K-coalgebra,
(ii) Δ_B and ϵ_B are homomorphisms of K-algebras.

The requirement that $\Delta_B : B \to B \otimes B$ be an algebra homomorphism implies that

$$\Delta_B(ab) = \sum_{(ab)} (ab)_{(1)} \otimes (ab)_{(2)}$$

$$= \Delta_B(a)\Delta_B(b)$$

$$= \left(\sum_{(a)} a_{(1)} \otimes a_{(2)} \right) \left(\sum_{(b)} b_{(1)} \otimes b_{(2)} \right)$$

$$= \sum_{(a,b)} a_{(1)}b_{(1)} \otimes a_{(2)}b_{(2)},$$

for $a, b \in B$.

Let B be a bialgebra. An element $b \in B$ for which $\Delta_B(b) = 1 \otimes b + b \otimes 1$ is a **primitive element** of B.

Example 2.1.2. Let K be a field, let S be a monoid. Then the monoid ring KS is a K-algebra $(KS, m_{KS}, \lambda_{KS})$ with multiplication map $m_{KS} : KS \otimes KS \to KS$ defined as $m_{KS}(a \otimes b) = ab$ and unit map $\lambda_A : K \to KS$ given as $\lambda_{KS}(r) = r$ for all $a, b \in KS$, $r \in K$.

Let $\Delta_{KS} : KS \to KS \otimes KS$ be the map defined as

$$\Delta_{KS}\left(\sum_{s \in S} r_s s \right) = \sum_{s \in S} r_s(s \otimes s),$$

and let $\epsilon_{KS} : KS \to K$ be the map defined by

$$\epsilon_{KS}\left(\sum_{s \in S} r_s s \right) = \sum_{s \in S} r_s.$$

Then $(KS, \Delta_{KS}, \epsilon_{KS})$ is a K-coalgebra. Moreover, as one can easily verify, Δ_{KS} and λ_{KS} are homomorphisms of K-algebras and so

$$(KS, \mathrm{m}_{KS}, \lambda_{KS}, \Delta_{KS}, \epsilon_{KS})$$

is a K-bialgebra called the **monoid bialgebra**.

Example 2.1.3. Let $K[x]$ be the K-algebra of polynomials in the indeterminate x. From Example 1.2.13, $K[x]$ has the structure of a coalgebra, with maps $\Delta_{K[x]}$, $\epsilon_{K[x]}$. Since the maps $\Delta_{K[x]}$ and $\epsilon_{K[x]}$ are K-algebra homomorphisms,

$$(K[x], \mathrm{m}_{K[x]}, \lambda_{K[x]}, \Delta_{K[x]}, \epsilon_{K[x]})$$

is a K-bialgebra. Note that $\Delta_{K[x]}(x) = x \otimes x$, thus this bialgebra is the **polynomial bialgebra with x grouplike**.

Example 2.1.4. Let $K[x]$ be the K-algebra of polynomials in the indeterminate x. From Example 1.2.14, $K[x]$ has the structure of a coalgebra, with maps $\Delta_{K[x]}$, $\epsilon_{K[x]}$. Since the maps $\Delta_{K[x]}$ and $\epsilon_{K[x]}$ are K-algebra homomorphisms,

$$(K[x], \mathrm{m}_{K[x]}, \lambda_{K[x]}, \Delta_{K[x]}, \epsilon_{K[x]})$$

is a K-bialgebra. Note that $\Delta_{K[x]}(x) = 1 \otimes x + x \otimes 1$, thus this bialgebra is the **polynomial bialgebra with x primitive**.

Let B be a K-bialgebra. A **biideal** I is a K-subspace of B that is both an ideal and a coideal.

Proposition 2.1.5. *Let $I \subseteq B$ be a biideal of B. Then B/I is a K-bialgebra.*

Proof. From Proposition 1.2.7, we have that B/I is a K-algebra. By Proposition 1.2.15, B/I is a K-coalgebra. One notes that $\Delta_{B/I}$ is an algebra map since Δ_B is an algebra map. Moreover, $\epsilon_{B/I}$ is an algebra map since that property holds for ϵ_B. □

Let B, B' be bialgebras. A K-linear map $\phi : B \to B'$ is a **bialgebra homomorphism** if ϕ is both an algebra and coalgebra homomorphism. The bialgebra homomorphism $\phi : B \to B'$ is an **isomorphism of bialgebras** if ϕ is a bijection.

Surprisingly, the bialgebra structures on $K[x]$ given in Examples 2.1.3 and 2.1.4 are the only bialgebra structures on $K[x]$ up to algebra isomorphism.

Proposition 2.1.6. *Suppose the polynomial algebra $K[x]$ is given the structure of a K-bialgebra. Then there is some $z \in K[x]$ so that $K[z] = K[x]$ and z is either grouplike or z is primitive.*

Proof. Let $K[x]$ be a bialgebra and suppose that

$$\Delta_{K[x]}(x) = \sum_{i=0}^{m} \sum_{j=0}^{n} b_{i,j} x^i \otimes x^j \in K[x] \otimes K[x],$$

for $b_{i,j} \in K$. Thus $\Delta_{K[x]}$ is a finite sum of tensors $b_{i,j}x^i \otimes x^j$ in which i is the degree of x in the left factor of the tensor and j is the degree of x in the right factor of the tensor. Let l denote the highest degree of x that occurs in the left factors of the tensors in the sum $\Delta_{K[x]}$. Then $b_{l,j} \neq 0$ for some j, $0 \leq j \leq n$; let j' be the maximal j for which $b_{l,j} \neq 0$.

Now,

$$(I_{K[x]} \otimes \Delta_{K[x]})\Delta_{K[x]}(x) \in K[x] \otimes K[x] \otimes K[x]$$

is a finite sum of tensors of the form $cx^i \otimes x^j \otimes x^k$, $c \in K$; i is the degree of x in the left-most factor in the tensor and k is the degree of x in the right-most factor of the tensor. Note that l is the highest degree of x that occurs in the left-most factors of the tensors in the sum $(I_{K[x]} \otimes \Delta_{K[x]})\Delta_{K[x]}(x)$.

Now,

$$(\Delta_{K[x]} \otimes I_{K[x]})\Delta_{K[x]}(x) = (\Delta_{K[x]} \otimes I_{K[x]})\left(\sum_{i=0}^{m} \sum_{j=0}^{n} b_{i,j}x^i \otimes x^j \right)$$

$$= \sum_{i=0}^{m} \sum_{j=0}^{n} b_{i,j}\Delta_{K[x]}(x^i) \otimes x^j$$

$$= \sum_{i=0}^{m} \sum_{j=0}^{n} b_{i,j}(\Delta_{K[x]}(x))^i \otimes x^j$$

since $\Delta_{K[x]}$ is an algebra homomorphism and so,

$$(\Delta_{K[x]} \otimes I_{K[x]})\Delta_{K[x]}(x) = \sum_{i=0}^{m} \sum_{j=0}^{n} b_{i,j}\left(\sum_{\alpha=0}^{m} \sum_{\beta=0}^{n} b_{\alpha,\beta}x^\alpha \otimes x^\beta \right)^i \otimes x^j$$

$$= b_{l,j'}^{l+1}x^{l^2} \otimes x^{lj'} \otimes x^{j'} + T,$$

where T is a sum of tensors in $K[x] \otimes K[x] \otimes K[x]$ of the form $cx^i \otimes x^j \otimes x^k$ with $i \leq l^2$. Since $b_{l,j'} \neq 0$, the highest power of x that occurs in the left-most factors of the tensors in the sum $(\Delta_{K[x]} \otimes I_{K[x]})\Delta_{K[x]}(x)$ is l^2. By the coassociative property of $\Delta_{K[x]}$, one has $l^2 = l$ and hence, either $l = 0$ or $l = 1$.

Now let r denote the highest degree of x that occurs in the right factors of the tensors in the sum $\Delta_{K[x]}(x)$. By the argument above applied to the right factors of the tensors, one concludes that either $r = 0$ or $r = 1$. Consequently,

$$\Delta_{K[x]}(x) = b_{0,0}(1 \otimes 1) + b_{0,1}(1 \otimes x) + b_{1,0}(x \otimes 1) + b_{1,1}(x \otimes x),$$

for $b_{0,0}, b_{0,1}, b_{1,0}, b_{1,1} \in K$.

Put $\Delta = \Delta_{K[x]}$, $\epsilon = \epsilon_{K[x]}$. Let $y = x - \epsilon(x)$ and let $\Delta(y) = \sum_{i=0}^{m} \sum_{j=0}^{n} a_{i,j} y^i \otimes y^j$. By comparing the leading coefficients in $(\Delta \otimes I)\Delta(y)$ and $(I \otimes \Delta)\Delta(y)$ as above, we conclude that $a_{i,j} = 0$ if $i > 1$ or $j > 1$. Since $\epsilon(y) = 0$, we also have $a_{0,0} = 0$ and $a_{0,1} = a_{1,0} = 1$. Thus

$$\Delta(y) = 1 \otimes y + y \otimes 1 + ay \otimes y$$

for some $a \in K$. If $a = 0$, then $z = y$ is primitive and $K[z] = K[x]$. If $a \neq 0$, put $z = 1 + ay$. Then z is group-like with $K[z] = K[x]$. $\square$

Let B be a bialgebra, and let A be an algebra and a left B-module with action denoted by "$\cdot$". Then A is a **left B-module algebra** if

$$b \cdot (aa') = \sum_{(b)} (b_{(1)} \cdot a)(b_{(2)} \cdot a')$$

and

$$b \cdot 1_A = \epsilon_B(b) 1_A$$

for all $a, a' \in A$, $b \in B$. Let A, A' be K-algebras. A K-linear map $\phi : A \to A'$ is a **left B-module algebra homomorphism** if ϕ is both an algebra and a left B-module homomorphism.

Let C be a coalgebra and a right B-module with action denoted by "$\cdot$". Then C is a **right B-module coalgebra** if

$$\Delta_C(c \cdot b) = \sum_{(c,b)} c_{(1)} \cdot b_{(1)} \otimes c_{(2)} \cdot b_{(2)}$$

and

$$\epsilon_C(c \cdot b) = \epsilon_C(c)\epsilon_B(b),$$

for all $c \in C$, $b \in B$. Let C, C' be K-coalgebras. A K-linear map $\phi : C \to C'$ is a **right B-module coalgebra homomorphism** if ϕ is both a coalgebra and a right B-module homomorphism.

Let B be a bialgebra. There is a left B-module structure on B^* defined as

$$(a \rightharpoonup f)(b) = f(ba),$$

for $a, b \in B$, $f \in B^*$. For $a \in B$, $f \in B^*$ the element $a \rightharpoonup f$ is the **left translate of f by a**. The left translate action endows B^* with the structure of a left B-module algebra: for $f, g \in B^*$, $a, b \in B$,

$$(a \rightharpoonup fg)(b) = (fg)(ba)$$

$$= m_{B^*}(f \otimes g)(ba)$$

$$= (f \otimes g)\Delta_B(ba)$$

$$= (f \otimes g)\left(\sum_{(b,a)} b_{(1)}a_{(1)} \otimes b_{(2)}a_{(2)} \right)$$

$$= \sum_{(b,a)} f(b_{(1)}a_{(1)})g(b_{(2)}a_{(2)})$$

$$= \sum_{(b,a)} (a_{(1)} \rightharpoonup f)(b_{(1)})(a_{(2)} \rightharpoonup g)(b_{(2)})$$

$$= \left(\sum_{(a)} (a_{(1)} \rightharpoonup f)(a_{(2)} \rightharpoonup g) \right)(b).$$

Moreover,

$$(a \rightharpoonup 1_{B^*})(b) = 1_{B^*}(ba) = \epsilon_B(ba) = \epsilon_B(b)\epsilon_B(a) = (\epsilon_B(a)1_{B^*})(b).$$

Likewise, there is a right B-module structure on B^* defined as

$$(f \leftharpoonup a)(b) = f(ab)$$

for all $a, b \in B, f \in B^*$. For $a \in B, f \in B^*$, the element $f \leftharpoonup a$ is the **right translate** of f by a. Note that $f \leftharpoonup B = \{f \leftharpoonup b : b \in B\}$ is a subspace of B^*.

For example, let $K[x]$ be the polynomial bialgebra with x grouplike. There is a right $K[x]$-module structure on $K[x]^*$ defined by

$$(f \leftharpoonup x^j)(x^k) = f(x^{j+k})$$

for $f \in K[x]^*$, $k, j \geq 0$. For instance, the right translate $e_i \leftharpoonup x^j$ (with e_i defined by $e_i(x^j) = \delta_{i,j}$) is defined as

$$(e_i \leftharpoonup x^j)(x^k) = e_i(x^{j+k}) = \delta_{i,j+k}.$$

Thus,

$$e_i \leftharpoonup x^j = \begin{cases} e_{i-j} & \text{if } i \geq j \\ 0 & \text{if } i < j. \end{cases}$$

Note that $e_i \leftharpoonup K[x]$ is the subspace of $K[x]^*$ generated by $\{e_0, e_1, e_2, \ldots, e_i\}$, hence $\dim(e_i \leftharpoonup K[x]) = i + 1$.

Lemma 2.1.7. *Let B be a K-bialgebra. Let $f \in B^*$. Then the following are equivalent.*

(i) $\dim(f \leftharpoonup B) < \infty$.
(ii) $f \in B^\circ$.

Proof. The proof is beyond the scope of this book. For a proof, the reader is referred to [Sw69], [Ab77, Lemma 2.2.2, Lemma 2.2.5], [Mo93, Lemma 9.1.1]. $\square$

We may not wish to give the proof here, but we can illustrate Lemma 2.1.7 (at least the (i) $\implies$ (ii) direction). As above, $\dim(e_i \leftharpoonup K[x]) = i + 1$, and so, $e_i \in K[x]^\circ$. Indeed, e_i corresponds to the $(i + 1)$st-order linearly recursive sequence in K with characteristic polynomial $f(x) = x^{i+1}$ and initial state vector $\underbrace{000\cdots 1}_{i+1}$; e_i vanishes on the ideal (x^{i+1}) of codimension $i + 1$.

Lemma 2.1.7 is the key to proving the following proposition.

Proposition 2.1.8. *If B is a bialgebra, then B° is a bialgebra.*

Proof. We first show that B° is an algebra; we need to construct a multiplication map m_{B° and a unit map λ_{B° that satisfy the associative and unit properties, respectively. By Proposition 1.3.1, B^* is an algebra with multiplication $m_{B^*} = \Delta_B^*$ and unit map $\lambda_{B^*} = \epsilon_B^*$. Let m_{B° denote the restriction of m_{B^*} to B°. We show that m_{B° : $B^\circ \otimes B^\circ \to B^\circ$. To this end, let $f, g \in B^\circ$ and let $a, b \in B$. Then

$$
\begin{aligned}
(fg \leftharpoonup a)(b) &= (m_{B^*}(f \otimes g) \leftharpoonup a)(b) \\
&= m_{B^*}(f \otimes g)(ab) \\
&= (f \otimes g)\Delta_B(ab) \\
&= (f \otimes g)\left(\sum_{(a),(b)} a_{(1)}b_{(1)} \otimes a_{(2)}b_{(2)} \right) \\
&= \sum_{(a,b)} f(a_{(1)}b_{(1)})g(a_{(2)}b_{(2)}) \\
&= \sum_{(a,b)} (f \leftharpoonup a_{(1)})(b_{(1)})(g \leftharpoonup a_{(2)})(b_{(2)}).
\end{aligned}
$$

Moreover,

$$
\begin{aligned}
\sum_{(a,b)} (f \leftharpoonup a_{(1)})(b_{(1)})(g \leftharpoonup a_{(2)})(b_{(2)}) &= \sum_{(a,b)} ((f \leftharpoonup a_{(1)}) \otimes (g \leftharpoonup a_{(2)}))(b_{(1)} \otimes b_{(2)}) \\
&= \sum_{(a)} ((f \leftharpoonup a_{(1)}) \otimes (g \leftharpoonup a_{(2)}))\left(\sum_{(b)} b_{(1)} \otimes b_{(2)} \right) \\
&= \sum_{(a)} ((f \leftharpoonup a_{(1)}) \otimes (g \leftharpoonup a_{(2)}))\Delta_B(b) \\
&= \sum_{(a)} m_{B^*}((f \leftharpoonup a_{(1)}) \otimes (g \leftharpoonup a_{(2)}))(b) \\
&= \left(\sum_{(a)} (f \leftharpoonup a_{(1)})(g \leftharpoonup a_{(2)}) \right)(b).
\end{aligned}
$$

Thus $fg \leftharpoonup B \subseteq \mathrm{span}((f \leftharpoonup B)(g \leftharpoonup B))$. Observe that

$$\dim(\mathrm{span}((f \leftharpoonup B)(g \leftharpoonup B))) < \infty$$

since $\dim(f \leftharpoonup B) < \infty$ and $\dim(g \leftharpoonup B) < \infty$ by Lemma 2.1.7. Consequently, $\dim(fg \leftharpoonup B) < \infty$. It follows that $fg \in B^\circ$ by Lemma 2.1.7. And so, m_{B° is a K-linear map $m_{B^\circ} : B^\circ \otimes B^\circ \to B^\circ$. Also, m_{B° satisfies the associative property since m_{B^*} does.

Next we need a unit map for B°. Our choice of course is $\lambda_{B^*} : K \to B^*$, but we need to show that $\lambda_{B^*}(K) \subseteq B^\circ$. Note that $\lambda_{B^*}(r) = r\lambda_{B^*}(1_K) = r1_{B^*} = r\epsilon_B$. Now, $\ker(\epsilon_B)$ is an ideal of B of finite codimension since the codomain of ϵ_B is K. Thus $\epsilon_B \in B^\circ$, and so $\lambda_{B^*} : K \to B^\circ$. Consequently, we take $\lambda_{B^\circ} = \lambda_{B^*}$. We conclude that $(B^\circ, m_{B^\circ}, \lambda_{B^\circ})$ is a K-algebra.

By Proposition 1.3.9, B° is a coalgebra with comultiplication map Δ_{B° and counit map ϵ_{B°. So to prove that B° is a bialgebra, it remains to show that Δ_{B° and ϵ_{B° are algebra homomorphisms. We consider comultiplication first. Let $f, g \in B^\circ$, $a, b \in B$. Then

$$\begin{aligned}
\Delta_{B^\circ}(fg)(a \otimes b) &= \Delta_{B^\circ}(m_{B^\circ}(f \otimes g))(a \otimes b) \\
&= m_{B^\circ}(f \otimes g)(m_B(a \otimes b)) \\
&= m_{B^\circ}(f \otimes g)(ab) \\
&= (f \otimes g)\Delta_B(ab) \\
&= (f \otimes g)\left(\sum_{(a,b)} a_{(1)}b_{(1)} \otimes a_{(2)}b_{(2)} \right) \\
&= \sum_{(a,b)} f(a_{(1)}b_{(1)})g(a_{(2)}b_{(2)}).
\end{aligned} \tag{2.1}$$

Now, $\displaystyle\sum_{(a,b)} f(a_{(1)}b_{(1)})g(a_{(2)}b_{(2)})$

$$\begin{aligned}
&= \sum_{(a,b)} f(m_B(a_{(1)} \otimes b_{(1)}))g(m_B(a_{(2)} \otimes b_{(2)})) \\
&= \sum_{(a),(b)} \Delta_{B^\circ}(f)(a_{(1)} \otimes b_{(1)})\Delta_{B^\circ}(g)(a_{(2)} \otimes b_{(2)}) \\
&= (\Delta_{B^\circ}(f) \otimes \Delta_{B^\circ}(g))(I \otimes \tau \otimes I)\left(\sum_{(a,b)} (a_{(1)} \otimes a_{(2)} \otimes b_{(1)} \otimes b_{(2)}) \right) \\
&= (\Delta_{B^\circ}(f) \otimes \Delta_{B^\circ}(g))(I \otimes \tau \otimes I)(\Delta_B \otimes \Delta_B)(a \otimes b) \\
&= (\Delta_{B^\circ}(f) \otimes \Delta_{B^\circ}(g))\Delta_{B \otimes B}(a \otimes b).
\end{aligned} \tag{2.2}$$

The reader should note that in moving from line 5 to line 6 above, we used the fact that the comultiplication of the coalgebra $B \otimes B$ is defined to be

$$\Delta_{B \otimes B} = (I_B \otimes \tau \otimes I_B)(\Delta_B \otimes \Delta_B).$$

The transpose of the map $\Delta_{B \otimes B}$ is

$$\Delta^*_{B \otimes B} : ((B \otimes B) \otimes (B \otimes B))^* \to (B \otimes B)^*$$

and $\Delta^*_{B \otimes B}$ restricted to $(B^\circ \otimes B^\circ) \otimes (B^\circ \otimes B^\circ)$ is the map

$$(\Delta^*_B \otimes \Delta^*_B)(I_{B^*} \otimes \tau \otimes I_{B^*}),$$

which is the multiplication on $B^\circ \otimes B^\circ$. Hence,

$$
\begin{aligned}
(\Delta_{B^\circ}(f) \otimes \Delta_{B^\circ}(g))\Delta_{B \otimes B}(a \otimes b) &= \Delta^*_{B \otimes B}(\Delta_{B^\circ}(f) \otimes \Delta_{B^\circ}(g))(a \otimes b) \\
&= m_{B^\circ \otimes B^\circ}(\Delta_{B^\circ}(f) \otimes \Delta_{B^\circ}(g))(a \otimes b) \\
&= (\Delta_{B^\circ}(f)\Delta_{B^\circ}(g))(a \otimes b). \qquad (2.3)
\end{aligned}
$$

From (2.1)–(2.3) we obtain

$$\Delta_{B^\circ}(fg) = \Delta_{B^\circ}(f)\Delta_{B^\circ}(g),$$

and so Δ_{B° is an algebra map. Next, we show that ϵ_{B° is an algebra map. For $r \in K$,

$$
\begin{aligned}
\epsilon_{B^\circ}(fg)(r) &= (fg)\lambda_B(r) \\
&= (m_{B^\circ}(f \otimes g))(\lambda_B(r1_K) \\
&= r(m_{B^\circ}(f \otimes g))(1_B) \\
&= r(f \otimes g)(\Delta_B(1_B)) \\
&= r(f \otimes g)(1_B \otimes 1_B) \\
&= rf(1_B)g(1_B).
\end{aligned}
$$

From the proof of Proposition 1.3.9, we have the formula

$$\epsilon_{B^\circ}(f) = f(1_B),$$

thus

$$
\begin{aligned}
rf(1_B)g(1_B) &= r\epsilon_{B^\circ}(f)\epsilon_{B^\circ}(g) \\
&= (\epsilon_{B^\circ}(f)\epsilon_{B^\circ}(g))(r).
\end{aligned}
$$

Consequently,

$$\epsilon_{B^\circ}(fg) = \epsilon_{B^\circ}(f)\epsilon_{B^\circ}(g)$$

and so ϵ_{B° is an algebra map. We conclude that B° is a bialgebra. $\square$

Proposition 2.1.9. *If B is cocommutative, then B° is a commutative. If B is a commutative, then B° is cocommutative.*

Proof. This is just a restatement of Proposition 1.3.11. $\square$

Proposition 2.1.10. *Suppose that B is a finite dimensional vector space over the field K. Then B is a bialgebra if and only if B^* is a bialgebra.*

Proof. Since B is finite dimensional, $B^\circ = B^*$. If B is an bialgebra, then B^* is a bialgebra by Proposition 2.1.8. Now, if B^* is a bialgebra, then $(B^*)^\circ$ is a bialgebra by Proposition 2.1.8. But since $\dim(B^*) = \dim(B) < \infty$, $(B^*)^\circ = B^{**}$ (the double dual) which is identified with B. Therefore B is a bialgebra. $\square$

As we have seen in §1.3, if $K[x]$ is the polynomial algebra, then the coalgebra $K[x]^\circ$ is the collection of linearly recursive sequences over K of all orders. But there are two coalgebra structures on $K[x]$ as given in Examples 2.1.3 and 2.1.4 making $K[x]$ into a bialgebra. (In fact, by Proposition 2.1.6 there are exactly two bialgebra structures on $K[x]$ up to algebra isomorphism.) And so, by Proposition 2.1.8 there are two bialgebra structures on $K[x]^\circ$.

This means that we can now multiply sequences in $K[x]^\circ$ in two different ways. If $K[x]$ is the polynomial bialgebra with x group-like, then $K[x]^\circ$ is the bialgebra with multiplication defined through the comultiplication on $K[x]$:

For $\{s_n\}, \{t_n\} \in K[x]^\circ$, $g(x) = \sum_{i=0}^{l} a_i x^i \in K[x]$,

$$(\{s_n\} \cdot \{t_n\})(g(x)) = (m_{K[x]^\circ}(\{s_n\} \otimes \{t_n\}))(g(x))$$

$$= (\{s_n\} \otimes \{t_n\})\Delta_{K[x]}(g(x))$$

$$= (\{s_n\} \otimes \{t_n\})\Delta_{K[x]}\left(\sum_{i=0}^{l} a_i x^i\right)$$

$$= (\{s_n\} \otimes \{t_n\})\sum_{i=0}^{l} a_i(x^i \otimes x^i)$$

$$= \sum_{i=0}^{l} a_i\{s_n\}(x^i)\{t_n\}(x^i)$$

$$= \sum_{i=0}^{l} a_i s_i t_i.$$

$$= \{s_n t_n\}(g(x)).$$

This is the **Hadamard product** of the sequences.

The Hadamard product $\{s_n\} \cdot \{t_n\} = \{s_n t_n\}$ is a linearly recursive sequence in K. How do we find its characteristic polynomial? We consider the situation in which both $\{s_n\}$ and $\{t_n\}$ are geometric sequences. Let $\{s_\blacksquare\}$ be the geometric sequence with characteristic polynomial $f(x) = x - \alpha$, $\alpha \in K$, and initial state vector $\mathbf{s}_0 = s_0$ and let $\{t_n\}$ be the geometric sequence with characteristic polynomial $g(x) = x - \beta$, $\beta \in K$ and initial state vector $\mathbf{t}_0 = t_0$. Then the Hadamard product is

$$\{s_n\} \cdot \{t_n\} = \{s_n t_n\}$$
$$= \{s_0 \alpha^n t_0 \beta^n\}$$
$$= \{s_0 t_0 (\alpha\beta)^n\},$$

which is a geometric sequence with characteristic polynomial $h(x) = x - \alpha\beta$ and initial state vector $s_0 t_0$. This is the essential idea behind the following proposition which we give without proof (see, [ZM73, CG93])

Proposition 2.1.11. *Let K be a field containing $\mathbb{Q}$. Let $\{s_n\}$ be a kth-order linearly recursive sequence in K with characteristic polynomial $f(x)$. Let $\{t_n\}$ be an lth-order linearly recursive sequence in K with characteristic polynomial $g(x)$. Suppose that $f(x), g(x)$ have distinct roots in some field extension L/K. Let $\alpha_1, \alpha_2, \ldots, \alpha_k$ be the distinct roots of $f(x)$ and let $\beta_1, \beta_2 \ldots, \beta_l$ be the distinct roots of $g(x)$. Then the characteristic polynomial of the Hadamard product $\{s_n\} \cdot \{t_n\} = \{s_n t_n\}$ is*

$$h(x) = \prod_{\substack{1 \le i \le k, \\ 1 \le j \le l, \\ \alpha_i \beta_j \ \text{distinct}}} (x - \alpha_i \beta_j).$$

Here is an illustration of Proposition 2.1.11.

Example 2.1.12. Let $\{s_n\}$ be the Fibonacci sequence in $\mathbb{Q}$ with characteristic polynomial $f(x) = x^2 - x - 1$ and initial state vector $\mathbf{s}_0 = (0, 1)$. The Hadamard product $\{s_n\} \cdot \{s_n\} = \{s_n^2\}$ is

$$0, 1, 1, 4, 9, 25, 64, 169, \ldots.$$

The zeros of $f(x)$ are $\alpha = \frac{1+\sqrt{5}}{2}$, $1 - \alpha = \frac{1-\sqrt{5}}{2}$, and so, the characteristic polynomial of the Hadamard product is

$$h(x) = (x - \alpha^2)(x - \alpha(1 - \alpha))(x - (1 - \alpha)^2)$$
$$= (x - \alpha^2)(x + 1)(x - (2 - \alpha))$$
$$= (x + 1)(x^2 - 3x + 1)$$
$$= x^3 - 2x^2 - 2x + 1.$$

Indeed, the Hadamard product is a 3rd-order linearly recursive sequence $\{r_n\}$ with recurrence relation

$$r_{n+3} = 2r_{n+2} + 2r_{n+1} - r_n$$

and initial state vector $\mathbf{r}_0 = (0, 1, 1)$.

If $K[x]$ is the polynomial bialgebra with x primitive, then $K[x]^\circ$ is the bialgebra with multiplication defined through the comultiplication on $K[x]$. This multiplication is called the **Hurwitz product** and is defined as follows. For $\{s_n\}, \{t_n\} \in K[x]^\circ$, $g(x) = \sum_{i=0}^{l} a_i x^i \in K[x]$,

$$
\begin{aligned}
(\{s_n\} * \{t_n\})(g(x)) &= (m_{[K[x]^\circ}(\{s_n\} \otimes \{t_n\}))(g(x)) \\
&= (\{s_n\} \otimes \{t_n\}) \Delta_{K[x]}(g(x)) \\
&= (\{s_n\} \otimes \{t_n\}) \Delta_{K[x]}\left(\sum_{i=0}^{l} a_i x^i \right) \\
&= (\{s_n\} \otimes \{t_n\}) \sum_{i=0}^{l} a_i \sum_{j=0}^{i} \binom{i}{j} (x^j \otimes x^{i-j}) \\
&= \sum_{i=0}^{l} a_i \sum_{j=0}^{i} \binom{i}{j} \{s_n\}(x^j)\{t_n\}(x^{i-j}) \\
&= \sum_{i=0}^{l} a_i \sum_{j=0}^{i} \binom{i}{j} s_j t_{i-j}. \\
&= \left\{ \sum_{j=0}^{n} \binom{n}{j} s_j t_{n-j} \right\} (g(x)).
\end{aligned}
$$

The Hurwitz product $\{s_n\} * \{t_n\} = \left\{ \sum_{j=0}^{n} \binom{n}{j} s_j t_{n-j} \right\}$ is a linearly recursive sequence in K. How do we find its characteristic polynomial? Again, we consider the case where $\{s_n\}$ and $\{t_n\}$ are geometric. One has

$$
\begin{aligned}
\{s_n\} * \{t_n\} &= \left\{ \sum_{j=0}^{n} \binom{n}{j} s_j t_{n-j} \right\} \\
&= \left\{ \sum_{j=0}^{n} \binom{n}{j} s_0 \alpha^j t_0 \beta^{n-j} \right\}
\end{aligned}
$$

$$= \left\{ s_0 t_0 \sum_{j=0}^{n} \binom{n}{j} \alpha^j \beta^{n-j} \right\}$$

$$= \{ s_0 t_0 (\alpha + \beta)^n \},$$

which is a geometric sequence with characteristic polynomial $h(x) = x - (\alpha + \beta)$ and initial state vector $s_0 t_0$. Here is the general result, stated without proof, but see [ZM73, CG93].

Proposition 2.1.13. *Let K be a field containing $\mathbb{Q}$. Let $\{s_n\}$ be a kth-order linearly recursive sequence in K with characteristic polynomial $f(x)$. Let $\{t_n\}$ be an lth-order linearly recursive sequence in K with characteristic polynomial $g(x)$. Suppose that $f(x), g(x)$ have distinct roots in some field extension L/K. Let $\alpha_1, \alpha_2, \ldots, \alpha_k$ be the distinct roots of $f(x)$ and let $\beta_1, \beta_2 \ldots, \beta_l$ be the distinct roots of $g(x)$. Then the characteristic polynomial of the Hurwitz product*

$$\{s_n\} * \{t_n\} = \left\{ \sum_{j=0}^{n} \binom{n}{j} s_j t_{n-j} \right\}$$

is

$$h(x) = \prod_{\substack{1 \leq i \leq k, \\ 1 \leq j \leq l, \\ \alpha_i + \beta_j \ \text{distinct}}} (x - (\alpha_i + \beta_j)).$$

Example 2.1.14. Let $\{s_n\}$ be the Fibonacci sequence in $\mathbb{Q}$ with characteristic polynomial $f(x) = x^2 - x - 1$ and initial state vector $s_0 = (0, 1)$. The Hurwitz product $\{s_n\} * \{s_n\}$ is

$$0, 0, 2, 6, 22, 70, 230, \ldots.$$

The zeros of $f(x)$ are $\alpha = \frac{1+\sqrt{5}}{2}$, $1-\alpha = \frac{1-\sqrt{5}}{2}$, and so, the characteristic polynomial of the Hurwitz product is

$$h(x) = (x - 2\alpha)(x - 1)(x - (2 - 2\alpha))$$

$$= x^3 - 3x^2 - 2x + 4.$$

Indeed, the Hurwitz product is a 3rd-order linearly recursive sequence $\{r_n\}$ with recurrence relation

$$r_{n+3} = 3r_{n+2} + 2r_{n+1} - 4r_n$$

and initial state vector $\mathbf{r}_0 = (0, 0, 2)$.

For further reading on the Hadamard and Hurwitz products of linearly recursive sequences from a Hopf algebra viewpoint, see [LT90] and [Ta94].

2.2   Myhill–Nerode Bialgebras

In this section we give an application of bialgebras to formal languages and finite
state machines (finite automata) of theoretical computer science. After introducing
finite automata, and giving some examples, we prove the Myhill–Nerode theorem
which tells us precisely when a language is accepted by a finite automaton. We
generalize the Myhill–Nerode theorem to an algebraic setting in which a certain
finite dimensional bialgebra (a Myhill–Nerode bialgebra) plays the role of the finite
automaton that accepts the language. We construct some examples of Myhill–
Nerode bialgebras. We see that a finite automaton determines a Myhill–Nerode
bialgebra, which in turn determines a (perhaps different!) finite automaton.

*   *   *

Let $\Sigma = \{a_1, a_2, \dots, a_k\}$ be a finite alphabet and let Σ^* denote the collection of
words of finite length formed from the letters in Σ. A word $x \in \Sigma^*$ is written as

$$x = a_{i_1} a_{i_2} \dots a_{i_m},$$

where $i_1, i_2, \dots, i_m$ is a finite sequence of integers in $\{1, 2, 3, \dots, k\}$. We assume that
Σ^* contains the empty word e of length 0. Also, Σ^* together with concatenation is
a monoid. A **language** is a subset $L \subseteq \Sigma^*$.

Loosely speaking, a finite automaton is a type of computing machine that reads
words in Σ^* and outputs a state. Here is a formal definition.

Definition 2.2.1. A **finite automaton** is a 5-tuple $M = (Q, \Sigma, \delta, q_0, F)$ consisting
of a finite set of **states** Q, an **input alphabet** Σ, a **transition function** $\delta : Q \times \Sigma \to
Q$, an **initial state** $q_0 \in Q$, and a set of **accepting states** $F \subseteq Q$.

The finite automaton starts in the initial state q_0. On the input $x = a_{i_1} a_{i_2} \dots a_{i_m} \in \Sigma^*$
the machine will move to a new state as follows. The machine reads the left-most
letter a_{i_1} and transitions to the state $q^{(1)} = \delta(q_0, a_{i_1})$. The machine then reads the
next letter a_{i_2} and moves to the state $q^{(2)} = \delta(q^{(1)}, a_{i_2})$. Next, the machine reads a_{i_3}
and moves to the state $q^{(3)} = \delta(q^{(2)}, a_{i_3})$, and so on. Eventually, the machine reads
the last letter a_{i_m} and halts at the final state $q^{(m)} = \delta(q^{(m-1)}, a_{i_m})$.

If the finite automaton M is in state q and halts in state q' upon reading input
$x \in \Sigma^*$, then we write $q' = \hat{\delta}(q, x)$. In the notation above,

$$\begin{aligned}
q^{(m)} &= \hat{\delta}(q_0, a_{i_1} a_{i_2} \cdots a_{i_m}) \\
&= \hat{\delta}(q^{(1)}, a_{i_2} a_{i_3} \cdots a_{i_m}) \\
&= \hat{\delta}(q^{(2)}, a_{i_3} a_{i_4} \cdots a_{i_m}), \\
&\ \ \vdots \\
&= \hat{\delta}(q^{(m-1)}, a_{i_m}).
\end{aligned}$$

Observe that $\hat{\delta}(q, a) = \delta(q, a)$ for $q \in Q$, $a \in \Sigma$.

A word $x \in \Sigma^*$ is **accepted** by the finite automaton M if on input x, when starting at initial state q_0, the finite automaton halts at a state in F. In other words, the automaton M accepts word x if and only if $\hat{\delta}(q_0, x) \in F$. A language L is **accepted** by M if the machine halts at an accepting state on all inputs $x \in L$.

Finite automata can be used to perform many important tasks. For instance, finite automata can be used to decide whether a given word in Σ^* has a certain property or not. A "parity-check" machine decides whether or not a given word in $\{0, 1\}^*$ has an even number of 1's.

Example 2.2.2 (Parity-Check Automaton). This machine decides whether or not a given word in $\{0, 1\}^*$ has an even number of 1's. We take $\Sigma = \{0, 1\}$ and define the finite automaton $M = (Q, \Sigma, \delta, q_0, F)$ as follows. The set of states of the automaton is $Q = \{q_0, q_1\}$, the initial state is q_0 and set of accepting states is $F = \{q_0\}$. The transition function $\delta : Q \times \Sigma \to Q$ is given by the table below.

	0	1
q_0	q_0	q_1
q_1	q_1	q_0

For example, on input 1001, reading from left to right, the Parity-Check Automaton reads 1 and moves from the initial state q_0 to the state $q_1 = \delta(q_0, 1)$. The machine then reads 0 and stays in the state $q_1 = \delta(q_1, 0)$, and again reads 0 to remain in the state $q_1 = \delta(q_1, 0)$, and finally reads 1 and halts at the final state $q_0 = \delta(q_1, 1)$. Since $q_0 \in F$, q_0 is an accepting state. The machine halted at an accepting state precisely because 1001 has an even number of 1's. On the other hand, on input 111 starting at state q_0, the machine reads 1 and moves to the state $q_1 = \delta(q_0, 1)$, reads 1 again and move to $q_0 = \delta(q_1, 1)$, and reads 1 a last time and halts at state $q_1 = \delta(q_0, 1)$, a non-accepting state. The machine halted at a non-accepting state because 111 has an odd number of 1's.

In general, on input $x \in \Sigma^*$, the Parity-Check Automaton will halt at q_0 if and only if x has an even number of 1's (x is accepted by the Parity-Check automaton). The language accepted by the Parity-Check automaton is precisely the set of words in $\{0, 1\}^*$ that have an even number of 1's. By accepting precisely those words which have an even number of 1's, the machine "decides" which words have an even number of 1's.

We can represent a finite automaton using a **state transition diagram**, which is a type of directed graph in which the vertices are the states and the directed edges, labeled with the letters of the alphabet, define the transition function. The transition $\delta(q_i, a) = q_j$ for states q_i, q_j and letter $a \in \Sigma$ is indicated by labeling with a the directed edge from q_i to q_j. In Figure 2.1 below, we give the state transition diagram for the Parity-Check automaton.

For example, we compute the halting state on input $x = 10011$ for the finite state diagram in Figure 2.1. Starting at the left arrow going into the initial state q_0, the machine reads digit 1 and moves to state q_1, it then reads the next two 0's and stays in state q_1, then the machine reads 1 and moves to state q_0, finally the last 1 sends the machine to the halting state q_1.

Fig. 2.1 Finite state diagram
for Parity-Check automaton.
Accepting state is q_0

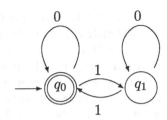

Fig. 2.2 Finite state diagram
for Check for a String Ending
in 11. Accepting state is q_2

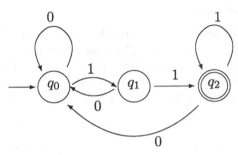

For our next example, we build a finite automaton that decides whether a given word in $\{0, 1\}^*$ ends in the string 11. The language accepted by this automaton will be the set of words in $\{0, 1\}^*$ that end in 11.

Example 2.2.3 (Check for a String Ending in 11). Let $M = (Q, \Sigma, \delta, q_0, F)$ with states $Q = \{q_0, q_1, q_2\}$, initial state q_0 and set of accepting states $F = \{q_2\}$. Define the transition function $\delta : Q \times \Sigma \to Q$ by the table:

	0	1
q_0	q_0	q_1
q_1	q_0	q_2
q_2	q_0	q_2

The state transition diagram for the automaton is given in Figure 2.2.

For instance, $x = 01011$ ends in 11, and so, on the input 01011 the machine halts in state q_2 (an accepting state), as required. On the other hand, on the input 101 the machine halts in state q_1 and so, 101 is not a word in the language accepted by M.

Here is another example of a finite automaton.

Example 2.2.4. We take $\Sigma = \{a, b\}$. Let $M = (Q, \Sigma, \delta, q_0, F)$ with states $Q = \{q_0, q_1\}$; the initial state is q_0 and the set of accepting states is $F = \{q_1\}$. The transition function is defined by

	a	b
q_0	q_1	q_0
q_1	q_1	q_1

The state transition diagram for the automaton is given in Figure 2.3.

Fig. 2.3 Finite state diagram
for automaton of
Example 2.2.4. Accepting
state is q_1

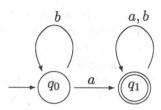

For this automaton, on the input *baab* the machine halts in state q_1, and so *baab* is accepted by M. On the other hand, *bb* is not accepted by this automaton. Can you describe the language accepted by M?

As we have seen, a given finite automaton $M = (Q, \Sigma, \delta, q_0, F)$ accepts the language L consisting of words $x \in \Sigma^*$ for which the machine halts in an accepting state on input x. But given an alphabet and a language $L \subseteq \Sigma^*$, does there exist a finite automaton that accepts L? This question has been settled by the famous Myhill–Nerode Theorem of theoretical computer science [Ei74, Chapter III, §9, Proposition 9.2], [HU79, §3.4, Theorem 3.9].

Let Σ be a finite alphabet and let $L \subseteq \Sigma^*$ be a language. On Σ^* we define an equivalence relation $\sim_L$ as follows: for $x, y \in \Sigma^*$,

$$x \sim_L y \text{ if and only if } xz \in L \text{ precisely when } yz \in L \text{ for all } z \in \Sigma^*.$$

The equivalence relation $\sim_L$ has **finite index** if there is a finite number of equivalence classes under $\sim_L$.

Proposition 2.2.5 (Myhill–Nerode Theorem). *Let Σ be a finite alphabet and let $L \subseteq \Sigma^*$ be a language. Then the following are equivalent.*

(i) The equivalence relation $\sim_L$ has finite index.
(ii) There exists a finite automaton that accepts language L.

Proof. *(i)* $\Longrightarrow$ *(ii)*. Suppose that $\sim_L$ has finite index. We need to construct a finite automaton which accepts L. For $x \in \Sigma^*$, let

$$[x] = \{y \in \Sigma^* : y \sim_L x\}$$

denote the equivalence class of $\sim_L$ containing x. Then $Q = \{[x] : x \in \Sigma^*\}$ is finite, and will constitute the states of the automaton.

Let $\delta : Q \times \Sigma \to Q$ be the relation defined as $\delta([x], a) = [xa]$ for $a \in \Sigma$. Suppose that $[x] = [y]$. Then $x \sim_L y$ and so, $x(az) \in L$ exactly when $y(az) \in L$ for all $z \in \Sigma^*$, whence $[xa] = [ya]$. Now,

$$\delta([x], a) = [xa] = [ya] = \delta([y], a)$$

and so δ is well-defined on classes in Q, in other words, $\delta : Q \times \Sigma \to Q$ is a function, which will serve as the transition function for our automaton. Setting $q_0 = [e]$ and $F = \{[x] \in Q : x \in L\}$ yields the finite automaton $(Q, \Sigma, \delta, q_0, F)$ that accepts L.

$(ii) \implies (i)$.   Suppose that L is accepted by the finite automaton $M = (Q, \Sigma, \delta, q_0, F)$. Let $\sim_M$ be the equivalence relation defined as $x \sim_M y$ if and only if $\hat{\delta}(q_0, x) = \hat{\delta}(q_0, y)$. Then $\sim_M$ has finite index since there are a finite number of states in Q. Note that

$$L = \{x \in \Sigma^* : \hat{\delta}(q_0, x) \in F\},$$

and so L is a union of some of the equivalence classes under $\sim_M$. Now suppose that $x \sim_M y$. Then $\hat{\delta}(q_0, x) = \hat{\delta}(q_0, y)$, and so, $\hat{\delta}(q_0, xz) = \hat{\delta}(q_0, yz)$ for all $z \in \Sigma^*$. Consequently, $xz \in L$ if and only if $yz \in L$ for all $z \in \Sigma^*$, that is, $x \sim_L y$. Thus the index of $\sim_L$ is less than or equal to the index of $\sim_M$. $\square$

To determine whether a given language L is accepted by a finite automaton, we compute the number of equivalences classes under $\sim_L$.

Example 2.2.6. Let L be the English language, which consists of words built from the alphabet $\Sigma = \{a, b, c, \ldots, z\}$. We claim that $\sim_L$ has finite index. We assume that L is finite, so that there is a word $x \in L$ of maximal length d. Now, any two words $x, y \in \Sigma^*$ of length $> d$ are equivalent since xz and yz are not in L for all $z \in \Sigma^*$. Thus $\sim_L$ admits at most $|L| + 1$ equivalence classes (the empty word e is not equivalent to any other word in Σ^*, and accounts for the additional class.) By the Myhill–Nerode Theorem the English language is accepted by a finite automaton.

Example 2.2.7. Let $L \subseteq \{a, b\}^*$ be defined as

$$L = \{xay : x, y \in \{a, b\}^*\}.$$

Now, $y \sim_L e$ if and only if $yz \in L$ precisely when $z \in L, \forall z \in \{a, b\}^*$. Thus $[e] = \{y : y \notin L\}$. Moreover, $y \sim_L a$ if and only if $yz \in L$ precisely when $az \in L, \forall z$. Since $a \in L$ and $e \notin L$, $[e] \neq [a]$, and so, $[a] = \{y : y \in L\}$. Thus the set of classes under $\sim_L$ is $\{[e], [a]\}$, with $L = [a]$.

By the Myhill–Nerode theorem, the language L is accepted by the finite automaton

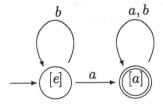

Example 2.2.8. Let $\Sigma = \{a, b\}$ and let L be the language consisting of all words in $\{a, b\}^*$ that contain no consecutive b's. One has $y \sim_L e$ if and only if $yz \in L$

precisely when $z \in L$, and so, $[e]$ consists of all $y \in L$ with the property that either $y = e$ or y ends in a. Now, $y \sim_L b$ if and only if $yz \in L$ precisely when $bz \in L$, and so, $[b]$ consists of all $y \in L$ with the property that y ends in b but not in bb. Finally, $y \sim_L bb$ if and only if $yz \in L$ precisely when $bbz \in L$, and so, $[bb]$ consists of all $y \notin L$. Consequently, the set of classes under $\sim_L$ is $\{[e], [b], [bb]\}$, with $L = [e] \cup [b]$.

By the Myhill–Nerode theorem, the language L is accepted by a finite automaton (see §2.4, Exercise 13).

Example 2.2.9. In this example, $\Sigma = \{a, b\}$ and $L = \{a^n b^n : n \geq 0\}$, that is,

$$L = \{e, ab, aabb, aaabbb, aaaabbbb, \dots\}.$$

As one can check, the set of classes under $\sim_L$ is

$$\{[b], [e], [a], [aa], [aaa], [aaaa], \dots\},$$

with $L = [e]$. Thus $\sim_L$ has infinite index, and so there is no finite automaton that accepts the language L.

For the moment we let K be a field, let S be any monoid, and let $H = KS$ be the monoid bialgebra with linear dual $H^* = KS^*$. There is a right H-module structure on H^* defined as

$$(p \leftharpoonup x)(y) = p(xy)$$

for all $x, y \in H, p \in H^*$. For $x \in H, p \in H^*$, the element $p \leftharpoonup x$ is the right translate of p by x; $p \leftharpoonup H = \{p \leftharpoonup x : x \in H\}$ is a subspace of H^*.

Now, let $S = \Sigma^*$ denote the monoid of words in a finite alphabet Σ and let $L \subseteq S$ be a language. Let $H = KS$ denote the monoid bialgebra. Let $\rho : S \to K$ be the **characteristic function** of L defined as

$$\rho(x) = \begin{cases} 1 & \text{if } x \in L \\ 0 & \text{if } x \notin L. \end{cases}$$

Then ρ extends by linearity to an element $p \in H^*$. Indeed,

$$p\left(\sum_{x \in S} a_x x\right) = \sum_{x \in S} a_x \rho(x).$$

Proposition 2.2.10. *Let $L \subseteq S$ be a language and let p be the element of $H^* = KS^*$ defined as above. Then $\sim_L$ has finite index if and only if the set of right translates $\{p \leftharpoonup x : x \in S\}$ is finite.*

Proof. Let $[x]$ denote the equivalence class of x under $\sim_L$. Define an equivalence relation $\sim_p$ on S as follows: $x \sim_p y$ if and only if $p(xz) = p(yz)$ for all $z \in S$. Then $\sim_L = \sim_p$. Let $[x]_p$ denote the equivalence class of x under $\sim_p$. Thus $\{[x] : x \in S\} = \{[x]_p : x \in S\}$.

Define a relation

$$\phi : \{[x]_p : x \in S\} \rightarrow \{p \leftarrow x : x \in S\}$$

by the rule $\phi([x]_p) = p \leftarrow x$. Suppose that $[x]_p = [y]_p$. Then

$$\phi([x]_p) = p \leftarrow x = p \leftarrow y = \phi([y]_p),$$

and so ϕ is well-defined on equivalence classes under $\sim_p$. Now assume that $\phi([x]_p) = \phi([y]_p)$. Then $p \leftarrow x = p \leftarrow y$, hence $[x]_p = [y]_p$, and so ϕ is an injection. Clearly, ϕ is surjective. The result follows. $\qquad\square$

Proposition 2.2.10 is the point of departure for generalizing the Myhill–Nerode theorem to an arbitrary monoid S: the condition "$\sim_L$ has finite index" gets replaced by the condition $\{p \leftarrow x : x \in S\}$ is finite. A certain bialgebra plays the role of the finite automaton that accepts the language, cf. [NU11].

Proposition 2.2.11 (Algebraic Myhill–Nerode Theorem). *Let S be a monoid and let $H = KS$ denote the monoid bialgebra. Let $p \in H^*$. Then the following are equivalent.*

(i) The set $\{p \leftarrow x : x \in S\}$ of right translates is finite.
(ii) There exists a finite dimensional bialgebra B, a bialgebra homomorphism $\Psi : H \rightarrow B$, and an element $f \in B^*$ so that $p(h) = f(\Psi(h))$ for all $h \in H$.

Proof. (i) $\Longrightarrow$ (ii). Let $Q = \{p \leftarrow x : x \in S\}$ be the finite set of right translates. For each $u \in S$, we define a right operator $r_u : Q \rightarrow Q$ by the rule

$$(p \leftarrow x)r_u = (p \leftarrow x) \leftarrow u = p \leftarrow xu.$$

Observe that the set $\{r_u : u \in S\}$ is finite with

$$|\{r_u : u \in S\}| \leq |Q|^{|Q|}.$$

We endow the set $\{r_u : u \in S\}$ with the binary operation "composition of operators" defined as follows: for $r_u, r_v \in \{r_u : u \in S\}, p \leftarrow x \in Q$,

$$(p \leftarrow x)(r_u r_v) = (p \leftarrow xu)r_v = p \leftarrow xuv = (p \leftarrow x)r_{uv}.$$

Thus $r_u r_v = r_{uv}$, for all $u, v \in S$. Then $\{r_u : u \in S\}$ together with composition of operators is a monoid with unity r_1.

Let B denote the monoid bialgebra on $\{r_u : u \in S\}$ over K. Let $\Psi : H \rightarrow B$ be the K-linear map defined by $\Psi(u) = r_u$. Then

$$\Psi(uv) = r_{uv} = r_u r_v = \Psi(u)\Psi(v).$$

Also,

$$\Delta_B(\Psi(u)) = \Delta_B(r_u)$$
$$= r_u \otimes r_u$$
$$= \Psi(u) \otimes \Psi(u)$$
$$= (\Psi \otimes \Psi)(u \otimes u)$$
$$= (\Psi \otimes \Psi)\Delta_H(u),$$

and

$$\epsilon_B(\Psi(u)) = \epsilon_B(r_u) = \epsilon_H(u),$$

and so, Ψ is a homomorphism of bialgebras.
Let $f \in B^*$ be defined by

$$f(r_u) = ((p \leftharpoonup 1)r_u)(1)$$
$$= (p \leftharpoonup u)(1)$$
$$= p(u)$$

Then $p(h) = f(\Psi(h))$, for all $h \in H$, as required.

$(ii) \implies (i)$.   Suppose there exists a finite dimensional bialgebra B, a bialgebra homomorphism $\Psi : H \to B$, and an element $f \in B^*$ so that $p(h) = f(\Psi(h))$ for all $h \in H$. Define a right H-module action $\cdot$ on B as

$$b \cdot h = b\Psi(h)$$

for all $b \in B, h \in H$. Then for $b \in B, x \in S$,

$$\Delta_B(b \cdot x) = \Delta_B(b\Psi(x))$$
$$= \Delta_B(b)\Delta_B(\Psi(x))$$
$$= \left(\sum_{(b)} b_{(1)} \otimes b_{(2)} \right)(\Psi \otimes \Psi)\Delta_H(x)$$
$$= \left(\sum_{(b)} b_{(1)} \otimes b_{(2)} \right)(\Psi(x) \otimes \Psi(x))$$
$$= \sum_{(b)} b_{(1)}\Psi(x) \otimes b_{(2)}\Psi(x)$$
$$= \sum_{(b)} b_{(1)} \cdot x \otimes b_{(2)} \cdot x$$

and

$$\epsilon_B(b \cdot x) = \epsilon_B(b\Psi(x)) = \epsilon_B(b)\epsilon_B(\Psi(x)) = \epsilon_B(b)\epsilon_H(x).$$

Thus B is a right H-module coalgebra.

Now, let Q be the collection of grouplike elements of B. By Proposition 1.2.18, Q is a linearly independent subset of B. Since B is finite dimensional, Q is finite. Since B is a right H-module coalgebra with action "$\cdot$",

$$\Delta_B(q \cdot x) = q \cdot x \otimes q \cdot x$$

for $q \in Q, x \in S$. Thus $\cdot$ restricts to give an action (also denoted by "$\cdot$") of S on Q. Now for $x, y \in S$,

$$
\begin{aligned}
(p \leftharpoonup x)(y) &= p(xy) \\
&= f(\Psi(xy)) \\
&= f(\Psi(x)\Psi(y)) \\
&= f((1_B\Psi(x))\Psi(y)) \\
&= f((1_B \cdot x) \cdot y) \tag{2.4}
\end{aligned}
$$

Let

$$T = \{q \in Q : q = 1_B \cdot x \text{ for some } x \in S\}$$

In view of (2.4) there exists a function

$$\varrho : T \to \{p \leftharpoonup x : x \in S\}$$

defined as

$$\varrho(1_B \cdot x)(y) = f((1_B \cdot x) \cdot y) = (p \leftharpoonup x)(y).$$

Since ϱ is surjective and T is finite, $\{p \leftharpoonup x : x \in S\}$ is finite. $\qquad\square$

The bialgebras constructed in Proposition 2.2.11(ii) are called **Myhill–Nerode bialgebras**. The easiest way to construct a Myhill–Nerode bialgebra is to start with a language $L \subseteq S$ for which $\sim_L$ has finite index. Then by Proposition 2.2.10, the set of right translates $\{p \leftharpoonup x : x \in S\}$ is finite, and so, by Proposition 2.2.11 (i)$\Longrightarrow$ (ii), there exists a Myhill–Nerode bialgebra B, a bialgebra homomorphism $\Psi : KS \to B$ and an element $f \in B^*$ so that $p(h) = f(\Psi(h))$, for all $h \in KS$. Here are some examples.

Example 2.2.12. Let $L \subseteq S = \Sigma^* = \{a, b\}^*$ be the language given in Example 2.2.7. In this case, the equivalence classes under $\sim_L$ are $[e]$ and $[a]$, with $L = [a]$. The characteristic function $\rho : S \to K$ extends to a function $p \in KS^*$, defined as $p\left(\sum_{x \in S} a_x x\right) = \sum_{x \in S} a_x \rho(x)$. There are exactly two right translates of p:

$$\{p \leftharpoonup 1, p \leftharpoonup a\}.$$

The set of right operators is $\{r_1, r_a\}$. The operators are defined by the table

$p \leftharpoonup x$	$(p \leftharpoonup x)r_1$	$(p \leftharpoonup x)r_a$
$p \leftharpoonup 1$	$p \leftharpoonup 1$	$p \leftharpoonup a$
$p \leftharpoonup a$	$p \leftharpoonup a$	$p \leftharpoonup a$

The monoid $T = \{r_1, r_a\}$ has binary operation given as

	r_1	r_a
r_1	r_1	r_a
r_a	r_a	r_a

The monoid bialgebra KT is the Myhill–Nerode bialgebra. Furthermore, the bialgebra homomorphism $\Psi : KS \to KT$ is given as $\Psi(x) = r_x$.

Let $\{e_{r_1}, e_{r_a}\}$ be the basis for KT^* dual to the basis $\{r_1, r_a\}$ for KT. Let $h = \sum_{x \in S} a_x x$. Then the element $e_{r_a} \in KT^*$ is so that

$$p(h) = p\left(\sum_{x \in S} a_x x\right) = \sum_{x \in S} a_x \rho(x)$$

$$= \sum_{x \in S} a_x \delta_{[a],[x]} = \sum_{x \in S} a_x e_{r_a}(r_x)$$

$$= e_{r_a}\left(\sum_{x \in S} a_x r_x\right) = e_{r_a}\left(\Psi\left(\sum_{x \in S} a_x x\right)\right)$$

$$= e_{r_a}(\Psi(h)),$$

as required. Also, KT^* is a K-bialgebra with $KT^* \cong K \times K$, as K-algebras and coalgebra structure defined as

$$\Delta_{KT^*}(e_{r_1}) = e_{r_1} \otimes e_{r_1},$$

$$\Delta_{KT^*}(e_{r_a}) = e_{r_1} \otimes e_{r_a} + e_{r_a} \otimes e_{r_1} + e_{r_a} \otimes e_{r_a},$$

$$\epsilon_{KT^*}(e_{r_1}) = e_{r_1}(r_1) = 1,$$

$$\epsilon_{KT^*}(e_{r_a}) = e_{r_a}(r_1) = 0.$$

Example 2.2.13. Let $\Sigma = \{a, b\}$ and let $L \subseteq \{a, b\}^*$ be the language consisting of all words that contain no consecutive b's (Example 2.2.8). The equivalences classes under $\sim_L$ are $[e], [b], [bb]$ with $L = [e] \cup [b]$. The characteristic function $\rho : S \to K$ extends to a function $p \in KS^*$, defined as $p\left(\sum_{x \in S} a_x x\right) = \sum_{x \in S} a_x \rho(x)$.

The set of right translates (states) is $Q = \{p \leftharpoonup 1, p \leftharpoonup b, p \leftharpoonup bb\}$. The set of right operators is $\{r_1, r_b, r_{bb}, r_a, r_{ab}, r_{ba}\}$. The operators are defined by the table

$p \leftharpoonup x$	$(p \leftharpoonup x)r_1$	$(p \leftharpoonup x)r_b$	$(p \leftharpoonup x)r_{bb}$	$(p \leftharpoonup x)r_a$	$(p \leftharpoonup x)r_{ab}$	$(p \leftharpoonup x)r_{ba}$
$p \leftharpoonup 1$	$p \leftharpoonup 1$	$p \leftharpoonup b$	$p \leftharpoonup bb$	$p \leftharpoonup 1$	$p \leftharpoonup b$	$p \leftharpoonup 1$
$p \leftharpoonup b$	$p \leftharpoonup b$	$p \leftharpoonup bb$	$p \leftharpoonup bb$	$p \leftharpoonup 1$	$p \leftharpoonup b$	$p \leftharpoonup bb$
$p \leftharpoonup bb$	$p \leftharpoonup bb$	$p \leftharpoonup bb$	$p \leftharpoonup bb$	$p \leftharpoonup bb$	$p \leftharpoonup bb$	$p \leftharpoonup bb$

The monoid $T = \{r_1, r_b, r_{bb}, r_a, r_{ab}, r_{ba}\}$ has binary operation given as

	r_1	r_b	r_{bb}	r_a	r_{ab}	r_{ba}
r_1	r_1	r_b	r_{bb}	r_a	r_{ab}	r_{ba}
r_b	r_b	r_{bb}	r_{bb}	r_{ba}	r_b	r_{bb}
r_{bb}	r_{bb}	r_{bb}	r_{bb}	r_{bb}	r_{bb}	r_{bb}
r_a	r_a	r_{ab}	r_{bb}	r_a	r_{ab}	r_a
r_{ab}	r_{ab}	r_{bb}	r_{bb}	r_a	r_{ab}	r_{bb}
r_{ba}	r_{ba}	r_b	r_{bb}	r_{ba}	r_b	r_{ba}

The monoid bialgebra KT is the Myhill–Nerode bialgebra and the bialgebra homomorphism $\Psi : KS \to KT$ is given as $\Psi(x) = r_x$. Let $\{e_{r_1}, e_{r_b}, e_{r_{bb}}, e_{r_a}, e_{r_{ab}}, e_{r_{ba}}\}$ be the basis for KT^* dual to the basis T for KT. Let $h = \sum_{x \in S} a_x x$. Then the element $1 - e_{r_{bb}} \in KT^*$ is so that

$$p(h) = p\left(\sum_{x \in S} a_x x \right) = \sum_{x \in S} a_x p(x)$$

$$= \sum_{x \in S} a_x \sum_{\substack{r_y \in T, \\ r_y \neq r_{bb}}} \delta_{[y],[x]} = \sum_{x \in S} a_x \sum_{\substack{r_y \in T, \\ r_y \neq r_{bb}}} e_{r_y}(r_x)$$

$$= \sum_{x \in S} a_x (1 - e_{r_{bb}})(r_x) = (1 - e_{r_{bb}})\left(\sum_{x \in S} a_x r_x \right)$$

$$= (1 - e_{r_{bb}})\left(\Psi\left(\sum_{x \in S} a_x x \right) \right) = (1 - e_{r_{bb}})(\Psi(h)),$$

as required.

Example 2.2.14. Let $\Sigma = \{a\}$, fix an integer $i \geq 0$ and let $L_i = \{a^i\} \subseteq S = \Sigma^*$, with characteristic function $\rho_i : S \to K$. Let $H = KS$ and let $p_i \in H^* = KS^*$ be the extension of ρ_i. The finite set of right translates of $p_i \in H^*$ is

$$Q_i = \{p_i \leftharpoonup 1, p_i \leftharpoonup a, p_i \leftharpoonup a^2, \ldots, p_i \leftharpoonup a^i, p_i \leftharpoonup a^{i+1}\}.$$

The set of right operators on Q_i is $T_i = \{r_1, r_a, r_{a^2}, \ldots, r_{a^i}, r_{a^{i+1}}\}$, a monoid. We have, for $0 \leq m, n \leq i + 1$,

$$r_{a^m} r_{a^n} = \begin{cases} r_{a^{m+n}} & \text{if } 0 \leq m + n \leq i + 1 \\ r_{a^{i+1}} & \text{if } m + n > i + 1 \end{cases}$$

$B_i = KT_i$ is the Myhill–Nerode bialgebra, with bialgebra homomorphism $\Psi : H \to B_i$ defined as $x \mapsto r_x$. The element $e_{r_{a^i}} \in B_i^*$ is so that

$$p_i(h) = e_{r_{a^i}}(\Psi(h)),$$

for all $h \in KS$.

Let B be a Myhill–Nerode bialgebra constructed from a language L with $\sim_L$ of finite index (as in Examples 2.2.12, 2.2.13, and 2.2.14).

Proposition 2.2.15. *The Myhill–Nerode bialgebra B determines a finite automaton $(Q, \Sigma, \delta, q_0, F)$ that accepts the language L.*

Proof. For the states of the automaton, we let Q be the (finite) set of group-like elements of B; this set is precisely the finite set of right operators $Q = \{r_x : x \in S\}$. For the input alphabet, we choose Σ. As we have seen, the right H-module structure of B restricts to an action "$\cdot$" of S on Q, and so we define the transition function $\delta : Q \times \Sigma \to Q$ by the rule

$$\delta(r_x, y) = r_x \Psi(y) = r_x r_y = r_{xy},$$

for $r_x \in Q$, $y \in S$. The initial state is $q_0 = 1_B$, and the set of final states F is the subset of Q of the form $1_B \cdot x$, $x \in S$ for which

$$p(x) = f(\Psi(x)) = f(1_B \Psi(x)) = f(1_B \cdot x) = 1$$

By construction, the finite automaton $(Q, \Sigma, \delta, 1_B, F)$ accepts L. $\qquad\square$

Example 2.2.16. Let B be the Myhill–Nerode bialgebra constructed in Example 2.2.12. The finite automaton is given as $(Q, \Sigma, \delta, 1_B, F)$ where $Q = \{r_1, r_a\}$, $F = \{r_a\}$, with transition function δ given by the table

	a	b
r_1	r_a	r_1
r_a	r_a	r_a

The finite automaton determined by the Myhill–Nerode bialgebra B is given in Figure 2.4 below.

The finite automaton given in Figure 2.4 accepts the language L of Example 2.2.12.

Example 2.2.17. Let B be the Myhill–Nerode bialgebra constructed in Example 2.2.13. The finite automaton is given as $(Q, \Sigma, \delta, 1_B, F)$ where $Q = \{r_1, r_b, r_{bb}, r_a, r_{ab}, r_{ba}\}$, $F = \{r_1, r_b, r_a, r_{ab}, r_{ba}\}$, with transition function δ given by the table

Fig. 2.4 Finite state diagram for Myhill–Nerode bialgebra KT. Accepting state is r_a

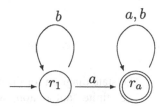

Fig. 2.5 Finite state diagram
for Myhill–Nerode bialgebra
KT. Accepting states are
$r_1, r_b, r_{ba}, r_a, r_{ab}$

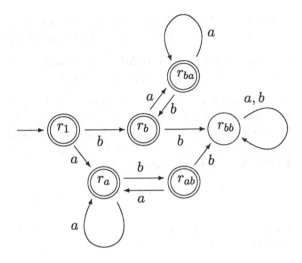

	a	b
r_1	r_a	r_b
r_b	r_{ba}	r_{bb}
r_{bb}	r_{bb}	r_{bb}
r_a	r_a	r_{ab}
r_{ab}	r_a	r_{bb}
r_{ba}	r_{ba}	r_b

The finite automaton associated with the Myhill–Nerode bialgebra B is given in Figure 2.5 above.

The finite automaton given in Figure 2.5 accepts the language L of Example 2.2.13.

Example 2.2.18. For $i \geq 0$, let B_i be the Myhill–Nerode bialgebra constructed in Example 2.2.14. The finite automaton is given as $(Q_i, \Sigma, \delta_i, 1_{B_i}, F_i)$ with $Q_i = \{r_1, r_a, r_{a^2}, \ldots, r_{a^i}, r_{a^{i+1}}\}$, $F = \{r_{a^i}\}$, with transition function δ_i given by the table

	a
r_1	r_a
r_a	r_{a^2}
r_{a^2}	r_{a^3}
$\vdots$	$\vdots$
r_{a^i}	$r_{a^{i+1}}$
$r_{a^{i+1}}$	$r_{a^{i+1}}$

The state diagram for the automaton is given below.

The finite automaton in Figure 2.6 accepts the language L given in Example 2.2.14.

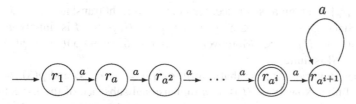

Fig. 2.6 Finite state diagram for B_i. Accepting state is r_{a^i}

2.3 Regular Sequences

In the final section of the chapter, we introduce regular sequences; these are sequences that generalize linearly recursive sequences over a Galois field.

$$* \quad * \quad *$$

Let S be a monoid which is countable as a set. Then the elements of S can be listed

$$x_0, x_1, x_2, x_3, \ldots.$$

Let K be a field, let $H = KS$ be the monoid bialgebra, and let $p \in H^*$. Define a sequence $\{s_n\}$ in K by the rule $s_n = p(x_n)$ for $n \geq 0$. Then $\{s_n\}$ is a **regular sequence** if the set of right translates $\{p \leftharpoonup x : x \in S\}$ is finite. Recall that for $x \in S$, the right translate $p \leftharpoonup x \in H^*$ is defined as $(p \leftharpoonup x)(y) = p(xy), \forall y \in S$.

Example 2.3.1. Let $S = \{a, b\}^*$ and let L be the language consisting of all words that contain no consecutive b's (see Example 2.2.8). The equivalence classes under $\sim_L$ are $[e]$, $[b]$, $[bb]$, and $L = [e] \cup [b]$. The set S is countable and its elements can be listed as

$$e, a, b, aa, ab, ba, bb, aaa, aab, aba, abb, baa, bab, bba, bbb, aaaa, aaab, \ldots$$

Here $x_0 = e$, $x_1 = a$, $x_2 = b$, and so on. Let $\rho : S \to K$ be the characteristic function of L, extending to an element $p = \sum_{n=0}^{\infty} \rho(x_n)e_n \in H^*$ with $e_n(x_m) = \delta_{n,m}$. The sequence $s_n = p(x_n)$ in K is

$$11111101110110011\ldots$$

The index of $\sim_L$ is 3 and so by Proposition 2.2.10, the set of right translates $\{p \leftharpoonup x : x \in S\}$ is finite. Thus $\{s_n\}$ is a regular sequence.

Proposition 2.3.2. *Assume the notation as above. Then $\{s_n\}$ is a regular sequence if and only if $\dim(p \leftharpoonup H) < \infty$ and the image $p(S) = \{p(x) : x \in S\}$ is finite.*

Proof. If $\{s_n\}$ is a regular sequence, then the set of right translates $\{p \leftharpoonup x : x \in S\}$ is finite. Since $\{p \leftharpoonup x : x \in S\}$ generates $p \leftharpoonup H$, $p \leftharpoonup H$ is finitely generated. Thus $\dim(p \leftharpoonup H) < \infty$. Moreover, for all $x \in S$, $(p \leftharpoonup x)(1) = p(x)$ and so, $\{p(x) : x \in S\}$ is finite.

For the converse, suppose that $\dim(p \leftharpoonup H) = n < \infty$ and $p(S) = \{p(x) : x \in S\}$ is finite. Now, $p \leftharpoonup H$ is an n-dimensional subspace of the K-vector space $\mathrm{Map}(S, K)$, and so, by a standard result there exists elements $y_1, y_2, \ldots, y_n \in S$ and a basis $\{f_1, f_2, \ldots, f_n\}$ for $p \leftharpoonup H$ for which $f_i(y_j) = \delta_{i,j}$ (see [NU11, Lemma 4.1]). Let $x \in S$. Then there exist elements $r_1, r_2, \ldots, r_n \in K$ for which $p \leftharpoonup x = r_1 f_1 + r_2 f_2 + \cdots + r_n f_n$. Evaluation at y_j yields $r_j = p(xy_j) \in p(S)$ for $1 \leq j \leq n$. Since $p(S)$ is finite, there are only a finite number of linear combinations of the f_i that represent right translates of p. Hence the number of right translates is finite and $\{s_n\}$ is regular. $\square$

As a consequence of Proposition 2.3.2 the notion of regular sequence does not depend on the order in which the elements of S are listed.

Proposition 2.3.3. *Let* $K = GF(p^m)$. *Let* $\{s_n\}$ *be a kth-order linearly recursive sequence in K. Then* $\{s_n\}$ *is a regular sequence.*

Proof. Let Σ be the alphabet consisting of the singleton letter $\{x\}$. Then $S = \Sigma^* = \{1, x, x^2, \ldots\}$ and $K[x]$ with the bialgebra structure of Example 2.1.3 is the monoid bialgebra $H = KS$. By Proposition 1.3.12 the sequence $\{s_n\}$ corresponds to an element $p = \sum_{n=0}^{\infty} s_n e_n$ in H° with $e_n(x^m) = \delta_{n,m}$. Clearly, $s_n = p(x^n)$ for all $n \geq 0$. Now by Lemma 2.1.7 $\dim(p \leftharpoonup H) < \infty$. Clearly, $p(S)$ is finite, and so by Proposition 2.3.2, $\{s_n\}$ is a regular sequence.

Alternatively, by the theory of linearly recursive sequences, $\{s_n\}$ is eventually periodic with period r, that is, there exist integers $N \geq 0$, $r > 0$ for which $s_{n+r} = s_n$ for all $n \geq N$. Consequently, the finite set

$$p \leftharpoonup 1, p \leftharpoonup x, p \leftharpoonup x^2, \ldots, p \leftharpoonup x^{N+r-1},$$

is a complete list of right translates of p. Thus $\{s_n\}$ is a regular sequence. $\square$

Example 2.3.4. Let $K = GF(2)$. Let $\{s_n\}$ be the 3rd-order linearly recursive sequence over K with characteristic polynomial $f(x) = x^3 + x + 1$, recurrence relation

$$s_{n+3} = s_{n+1} + s_n, \ n \geq 0,$$

and initial state vector $s_0 = 101$. Since $f(x)$ is a primitive polynomial over K (proof?), $\{s_n\}$ is periodic with period $r = 2^3 - 1 = 7$. Indeed, the sequence is

$$10111001011100\ldots,$$

which is regular by Proposition 2.3.3.

Let $\{s_n\}$ be the regular sequence in $K = GF(2)$ of Example 2.3.4 with $p = \sum_{n=0}^{\infty} s_n e_n \in K[x]^{\circ}$. The finite set of right translates is

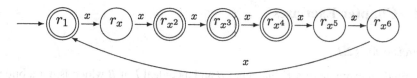

Fig. 2.7 Finite state diagram accepting language L. Accepting states are $r_1, r_{x^2}, r_{x^3}, r_{x^4}$

$$\{p \leftharpoonup 1, p \leftharpoonup x, p \leftharpoonup x^2, p \leftharpoonup x^3, p \leftharpoonup x^4, p \leftharpoonup x^5, p \leftharpoonup x^6\}.$$

Consequently, Proposition 2.2.11 applies to show that there is a Myhill–Nerode bialgebra B, a bialgebra homomorphism $\Psi : K[x] \to B$ and an element $f \in B^*$ for which $p(h) = f(\Psi(h))$ for all $h \in K[x]$. We construct B as follows. The right operators are

$$T = \{r_1, r_x, r_{x^2}, r_{x^3}, r_{x^4}, r_{x^5}, r_{x^6}\},$$

with binary operation given as

	r_1	r_x	r_{x^2}	r_{x^3}	r_{x^4}	r_{x^5}	r_{x^6}
r_1	r_1	r_x	r_{x^2}	r_{x^3}	r_{x^4}	r_{x^5}	r_{x^6}
r_x	r_x	r_{x^2}	r_{x^3}	r_{x^4}	r_{x^5}	r_{x^6}	r_1
r_{x^2}	r_{x^2}	r_{x^3}	r_{x^4}	r_{x^5}	r_{x^6}	r_1	r_x
r_{x^3}	r_{x^3}	r_{x^4}	r_{x^5}	r_{x^6}	r_1	r_x	r_{x^2}
r_{x^4}	r_{x^4}	r_{x^5}	r_{x^6}	r_1	r_x	r_{x^2}	r_{x^3}
r_{x^5}	r_{x^5}	r_{x^6}	r_1	r_x	r_{x^2}	r_{x^3}	r_{x^4}
r_{x^6}	r_{x^6}	r_1	r_x	r_{x^2}	r_{x^3}	r_{x^4}	r_{x^5}

The Myhill–Nerode bialgebra B is $KT \cong KC_7$. Note that KT^* has basis

$$\{e_{r_1}, e_{r_x}, e_{r_{x^2}}, e_{r_{x^3}}, e_{r_{x^4}}, e_{r_{x^5}}, e_{r_{x^6}}\}$$

with $e_{r_m}(r_{x^n}) = \delta_{m,n}$. There exists a bialgebra homomorphism $\Psi : K[x] \to KT$ defined as $\Psi(x) = r_x$. Let

$$f = e_{r_1} + e_{r_{x^2}} + e_{r_{x^3}} + e_{r_{x^4}}.$$

Then $p(h) = f(\Psi(h)), \forall h \in K[x]$, as required.

In addition, the sequence $\{s_n\}$ determines the language $L \subseteq \{x\}^*$ defined as: for $n \geq 0$, $x^n \in L$ if and only if $s_n = 1$. The language L is accepted by the finite automaton of Figure 2.7.

2.4   Chapter Exercises

Exercises for §2.1

1. Find an example of a K-bialgebra B and a coideal I of B which is not a biideal of B.

2. Let B be a commutative K-bialgebra. Show that $B \cong K[\{x_\alpha\}]/N$, as K-algebras, for some set of indeterminates $\{x_\alpha\}$, and some ideal N of $K[\{x_\alpha\}]$. Under what conditions does this isomorphism extend to an isomorphism of bialgebras?

3. Let B be a bialgebra and let $f \in B^*$. Prove that $f \leftharpoonup B$ is a subspace of B^*.

4. Let B be a bialgebra and let $f \in B^\circ$. Prove that $f \leftharpoonup B$ is a finite dimensional subspace of B^* (see §2.4, Exercise 3).

5. Let B be a K-bialgebra and let C be a right B-module coalgebra. Show that C^* is a left B-module algebra.

6. Let $\{s_n\}$ be the 2nd-order linearly recursive sequence in $\mathbb{Q}$ with recurrence relation $s_{n+2} = s_n$ and initial state vector $\mathbf{s}_0 = (1, 1)$. Let $\{t_n\}$ be the 1st-order linearly recursive sequence in $\mathbb{Q}$ with recurrence relation $t_{n+1} = 2t_n$ and $\mathbf{t}_0 = (1)$.

   (a) Compute the first 5 terms of the Hadamard product $\{s_n\} \cdot \{t_n\}$.
   (b) Compute the characteristic polynomial of the Hadamard product $\{s_n\} \cdot \{t_n\}$.

7. Let $\{s_n\}$ be the 4th-order linearly recursive sequence in $\mathbb{Q}$ with recurrence relation

$$s_{n+4} = -s_{n+3} - s_{n+2} - s_{n+1} - s_n$$

   and initial state vector $\mathbf{s}_0 = (2, 1, 0, 3)$.

   (a) Compute the first 5 terms of the Hadamard product $\{s_n\} \cdot \{s_n\}$.
   (b) Compute the characteristic polynomial of the Hadamard product $\{s_n\} \cdot \{s_n\}$.

8. Let $\{s_n\}$ be the 2nd-order linearly recursive sequence in $\mathbb{Q}$ with recurrence relation

$$s_{n+2} = -s_{n+1} - s_n$$

   and initial state vector $\mathbf{s}_0 = (1, 2)$.

   (a) Compute the first 5 terms of the Hurwitz product $\{s_n\} * \{s_n\}$.
   (b) Find the recurrence relation and the initial state vector of the Hurwitz product $\{s_n\} * \{s_n\}$.

Exercises for §2.2

9. Let $M = (Q, \Sigma, \delta, q_0, F)$ be the finite automaton with states $Q = \{q_0, q_1, q_2\}$, input alphabet $\Sigma = \{0, 1\}$, initial state q_0, set of accepting states $F = \{q_1, q_2\}$ and transition function $\delta : Q \times \Sigma \to Q$ given by the table:

	0	1
q_0	q_1	q_1
q_1	q_0	q_2
q_2	q_2	q_0

(a) Construct the state transition diagram for M.

(b) Compute $\hat{\delta}(q_0, 01101)$; $\hat{\delta}(q_1, 1001)$.

(c) Give a description of all words $x \in \{0, 1\}^*$ for which $\hat{\delta}(q_0, x) = q_2$.

10. Let Σ_0 be a finite alphabet and let $L \subseteq \Sigma_0^*$ be a language. Prove that $\sim_L$ is an equivalence relation.

11. Let L be the language accepted by the Parity-Check Automaton of Example 2.2.2. Compute the equivalence classes under $\sim_L$.

12. Let L be the language accepted by the Check for a String Ending in 11 automaton of Example 2.2.3. Compute the equivalence classes under $\sim_L$.

13. Construct a finite automaton that accepts the language L of Example 2.2.8. Does your finite automaton coincide with the finite automaton given in Example 2.2.17?

14. Let $M = (Q, \Sigma, \delta, q_0, F)$ be a finite automaton and let $\sim_M$ be the relation on Σ^* defined as $x \sim_M y$ if and only if $\hat{\delta}(q_0, x) = \hat{\delta}(q_0, y)$. Prove that $\sim_M$ is an equivalence relation.

15. Suppose the language L is accepted by a finite automaton. Can the "only if" part of the Myhill–Nerode theorem be proved using Proposition 2.2.11 (ii) $\implies$ (i)?

16. Construct the Myhill–Nerode bialgebra associated with the finite automaton of Example 2.2.2.

17. Construct the Myhill–Nerode bialgebra associated with the finite automaton of Example 2.2.3.

18. Let K be a field, let G be an infinite group, let $g \in G$ and let $e_g \in KG^*$ be defined as $e_g(h) = \delta_{g,h}$. Prove that e_g has an infinite number of right translates.

19. Let S be a monoid, let $H = KS$ be the monoid bialgebra and let $p \in H^*$. Suppose there exists a finite dimensional bialgebra B, a bialgebra homomorphism $\Psi : H \to B$ and an element $f \in B^*$ for which $p(h) = f(\Psi(h))$, $\forall h \in H$. Let $\phi : B \to A$ be an isomorphism of bialgebras. Prove that there exists a bialgebra homomorphism Ψ' and an element $g \in A^*$ for which $p(h) = g(\Psi'(h))$, $\forall h \in H$.

20. Let S be a monoid, let $H = KS$ be the monoid bialgebra and let $p \in H^*$. Suppose there exists a finite dimensional bialgebra B, a bialgebra homomorphism $\Psi : H \to B$ and an element $f \in B^*$ for which $p(h) = f(\Psi(h))$, $\forall h \in H$. Let $D \to H$ be an isomorphism of bialgebras. Show there exists an element $q \in D^*$ and a bialgebra homomorphism $\Psi' : D \to B$ for which $q(d) = f(\Psi'(d))$, $\forall d \in D$.

21. Let K be a field, let S be a monoid, let $p \in KS^*$ with $p(s) \in \{0, 1\}$, $\forall s \in S$. Prove the following: $p \in KS^\circ$ if and only if there exists a finite dimensional bialgebra B, a bialgebra homomorphism $\Psi : KS \to B$ and an element $f \in B^*$ for which $p(h) = f(\Psi(h))$, $\forall h \in KS$. In this sense, B "accepts" p; $L = \{s : p(s) = 1\}$ is the language accepted by B.

Exercises for §2.3

22. Let $\{s_n\}$ be a kth-order linearly recursive sequence over $GF(p^m)$ with characteristic polynomial $f(x) \in GF(p^m)[x]$. By Proposition 2.3.3, $\{s_n\}$ is a regular sequence. In this sense regular sequences generalize linearly recursive sequences. What is the analog of the characteristic polynomial for regular sequences?

Questions for Further Study

1. Let M be the finite automaton given by the state transition diagram below. (Note that the accepting states are q_1, q_3.)

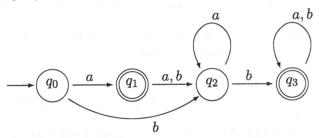

   (a) Compute the Myhill–Nerode bialgebra B associated to the automaton M.
   (b) Let $S = \{a, b\}^*$, let KS be the monoid bialgebra and let $p \in KS^*$ be the functional corresponding to the language L accepted by M. Find $f \in B^*$ so that $p(h) = f(\Psi(h))$ for all $h \in KS$.

2. Let M and M' be finite state machines which accept the languages L and L', respectively. Suppose M determines Myhill–Nerode bialgebra B and suppose M' determines Myhill–Nerode bialgebra B'. If B is isomorphic to B', are the languages L, L' the same?

3. Can every monoid bialgebra KS, with S finite, be viewed as a Myhill–Nerode bialgebra associated with some finite automaton (or equivalently, some language L with $\sim_L$ of finite index)?

Chapter 3
Hopf Algebras

In this chapter we introduce the notion of a Hopf algebra over a field K as a bialgebra with an additional map called the coinverse (or antipode). We discuss some basic features of Hopf algebras and give some initial examples, including the group ring KG. In many respects, KG is the example that is generalized in the concept of Hopf algebra. The coinverse map σ_{KG} of KG has order 2 under function composition, though this is not the case for every Hopf algebra. If H is a cocommutative Hopf algebra, however, then the coinverse σ_H has order 2, and so, cocommutative Hopf algebras generalize the group ring Hopf algebra KG.

We next consider certain subspaces of a Hopf algebra H, called the space of left (or right) integrals. The left integrals $\int_H^l$ can be described as the set of elements of H for which the action of H by left multiplication is trivial.

We then generalize the comultiplication map Δ_H to yield the concept of a right comodule over H and show that H^* is a right H-comodule. This leads to the concept of a Hopf module over H—a vector space that is both an H-module and an H-comodule. We state the Fundamental Theorem of Hopf Modules, a curious result which says that every right Hopf module over a finite dimensional Hopf algebra H is isomorphic to a trivial right Hopf module. We show that the right H-comodule H^* is a right Hopf module and prove a special case of the Fundamental Theorem which says that $H^* \cong \int_{H^*}^l \otimes H$, where $\int_{H^*}^l \otimes H$ has the structure of a trivial right H-Hopf module. As a corollary, we conclude that $\int_{H^*}^l$ is a one-dimensional subspace of H^*, and hence H^* admits a generating integral.

The last two sections of the chapter concern Hopf algebras over rings. Many properties of Hopf algebras over fields carry over to rings, including the notion of integrals and the Fundamental Theorem of Hopf Modules. The resulting iso-morphism $H^* \cong \int_{H^*}^l \otimes H$, valid for Hopf algebras over rings of finite rank, has been used in the classification of Hopf algebra orders in KC_{p^2}, see [Un94], [Un11, Chapter 8].

Next, we introduce Hopf orders in the group ring Hopf algebra KG. Hopf orders are Hopf algebras over rings which in some sense generalize the notion of fractional

© Springer International Publishing Switzerland 2015
R.G. Underwood, *Fundamentals of Hopf Algebras*, Universitext,
DOI 10.1007/978-3-319-18991-8_3

ideals over R in K where R is a Dedekind domain. When R is the ring of integers of a finite extension $K/\mathbb{Q}$, we construct a collection of one parameter Hopf orders $H(i)$ in KC_p together with their dual modules, $H(i)^D$. In §4.5, we will show how Hopf orders can be realized as "Galois groups" in a sense that will be made precise.

3.1   Introduction to Hopf Algebras

In this section we define a K-Hopf algebra H as a K-bialgebra with an additional map called the coinverse (or antipode) satisfying the coinverse (or antipode) property. We give some examples of Hopf algebras including the group ring KG which in many respects is the prototype Hopf algebra that others are modelled on. We define convolution, a binary operation on linear transformations, and use it to prove that the coinverse is an algebra anti-homomorphism, and consequently, that the coinverse has order 2 under composition whenever H is cocommutative. Using convolution, we also show that the coinverse is a coalgebra anti-homomorphism, a property that can be used to show that the coinverse has order 2 whenever H is commutative. (We will again employ the coalgebra anti-homomorphism property in §3.2.) Next, we consider Hopf ideals, quotient Hopf algebras, and homomorphisms of Hopf algebras. Finally we show that H is a finite dimensional Hopf algebra if and only if H^* is a finite dimensional Hopf algebra.

$$* \quad * \quad *$$

Definition 3.1.1. A **K-Hopf algebra** is a bialgebra over a field K

$$H = (H, m_H, \lambda_H, \Delta_H, \epsilon_H)$$

together with a K-linear map $\sigma_H : H \to H$ that satisfies

$$m_H(I_H \otimes \sigma_H)\Delta_H(h) = \epsilon_H(h)1_H = m_H(\sigma_H \otimes I_H)\Delta_H(h) \qquad (3.1)$$

for all $h \in H$. The map σ_H is the **coinverse** (or **antipode**) map and property (3.1) is the **coinverse** (or **antipode**) *property*.

The field K itself is a K-Hopf algebra (take $\sigma_K = I_K$) called the **trivial K-Hopf algebra**. Here are some other examples of Hopf algebras.

Example 3.1.2. Let G be a finite group. Let KG be the monoid (group) bialgebra of Example 2.1.2. Define a coinverse map $\sigma_{KG} : KG \to KG$ by

$$\sigma_{KG}(\tau) = \tau^{-1},$$

for $\tau \in G$. Then KG is a K-Hopf algebra.

Example 3.1.3. Let $K[x]$ be the polynomial bialgebra with x primitive (Example 2.1.4.) Define the coinverse map $\sigma_{K[x]} : K[x] \to K[x]$ by

$$\sigma_{K[x]}(x^i) = (-x)^i,$$

for $i \geq 0$. Then $K[x]$ is a K-Hopf algebra.

The polynomial bialgebra of Example 2.1.3 with x grouplike, that is, where $x \mapsto x \otimes x$ under comultiplication, cannot be endowed with the structure of a K-Hopf algebra, see §3.5, Exercise 1. We can modify this bialgebra to yield a Hopf algebra, however.

Example 3.1.4. Let $K[x,y]/(xy-1)$ be the quotient K-algebra. Then

$$K[x,y]/(xy-1) \cong K[x,x^{-1}].$$

Define comultiplication

$$\Delta_{K[x,x^{-1}]} : K[x,x^{-1}] \to K[x,x^{-1}] \otimes K[x,x^{-1}]$$

by making x and x^{-1} grouplike. Define the counit map

$$\epsilon_{K[x,x^{-1}]} : K[x,x^{-1}] \to K$$

by $x \mapsto 1, x^{-1} \mapsto 1$, and define the coinverse map

$$\sigma_{K[x,x^{-1}]} : K[x,x^{-1}] \to K[x,x^{-1}]$$

by $x \mapsto x^{-1}, x^{-1} \mapsto x$. Then $K[x,x^{-1}]$ is a K-Hopf algebra.

The next example is due to M. Sweedler, cf. [Mo93, 1.5.6].

Example 3.1.5. Let K be a field of characteristic $\neq 2$. Let H be the K-algebra generated by $\{1, g, x, gx\}$ modulo the relations

$$g^2 = 1, \; x^2 = 0, \; xg = -gx.$$

Let comultiplication $\Delta_H : H \to H \otimes_K H$ be defined by

$$g \mapsto g \otimes g, \; x \mapsto x \otimes 1 + g \otimes x,$$

let the counit map $\epsilon_H : H \to K$ be defined as $g \mapsto 1, x \mapsto 0$, and let the coinverse map $\sigma_H : H \to H$, be given by $g \mapsto g, x \mapsto -gx$. Then H is a K-Hopf algebra.

A K-Hopf algebra H is **commutative** if it is a commutative algebra; H is **cocommutative** if it is a cocommutative coalgebra. The group algebra KG of Example 3.1.2 is cocommutative; it is commutative if and only if G is abelian.

The Hopf algebras of Examples 3.1.3 and 3.1.4 are both commutative and cocommutative, while Sweedler's Hopf algebra of Example 3.1.5 is neither commutative nor cocommutative. A K-Hopf algebra that is neither commutative nor cocommutative is a **quantum group**.

In many ways, the group ring KG of Example 3.1.2 is the canonical example that is generalized to the concept of Hopf algebra. One has

$$\sigma_{KG} \circ \sigma_{KG} = I_{KG},$$

so that the coninverse of KG has order 2. We wonder: does every Hopf algebra have coinverse of order 2? The answer is "no." Consider the Sweedler Hopf algebra H of Example 3.1.5. We have

$$(\sigma_H \circ \sigma_H)(x) = \sigma_H(\sigma_H(x))$$
$$= \sigma_H(-gx)$$
$$= -x,$$

and so, σ_H does not have order 2. (In fact, σ_H has order 4, see §3.5, Exercise 4.)

What properties of a Hopf algebra H guarantee that its coinverse map σ_H has order 2, as does the coinverse σ_{KG} of the prototypical Hopf algebra KG? We shall answer this question in what follows. For the moment we let C be a K-coalgebra and let A be a K-algebra. Let $\text{Hom}_K(C,A)$ denote the collection of linear transformations $\phi : C \to A$. On $\text{Hom}_K(C,A)$ we can define a multiplication as follows. For $f, g \in \text{Hom}_K(C,A)$, $a \in C$,

$$(f * g)(a) = m_A(f \otimes g)\Delta_C(a) = \sum_{(a)} f(a_{(1)})g(a_{(2)}).$$

This multiplication is called **convolution**.

Proposition 3.1.6. *Let C be a K-coalgebra and let A be a K-algebra. Then $\text{Hom}_K(C,A)$ together with convolution $*$ is a monoid.*

Proof. We show that the axioms for a monoid hold. We first show that $*$ is associative. For $f, g, h \in \text{Hom}_K(C,A)$, one has

$$(f * (g * h))(a) = m_A(f \otimes (g * h))\Delta_C(a)$$
$$= \sum_{(a)} f(a_{(1)})(g * h)(a_{(2)})$$
$$= \sum_{(a)} f(a_{(1)}) \sum_{(a_{(2)})} g(a_{(2)(1)})h(a_{(2)(2)}),$$

which in Sweedler notation equals $\sum_{(a)} f(a_{(1)})g(a_{(2)})h(a_{(3)})$. Now, by the coassociativity of Δ_C,

$$\sum_{(a)} f(a_{(1)})g(a_{(2)})h(a_{(3)}) = \sum_{(a)} \sum_{(a_{(1)})} f(a_{(1)_{(1)}})g(a_{(1)_{(2)}})h(a_{(2)})$$

$$= \sum_{(a)} (f * g)(a_{(1)})h(a_{(2)})$$

$$= m_A((f * g) \otimes h)\Delta_C(a)$$

$$= ((f * g) * h)(a),$$

and so $*$ is associative.

Next, we show that $\lambda_A \epsilon_C$ serves as a left and right identity element in $\mathrm{Hom}_K(C, A)$. For $\phi \in \mathrm{Hom}_K(C, A)$, $a \in C$,

$$(\lambda_A \epsilon_C * \phi)(a) = m_A(\lambda_A \epsilon_C \otimes \phi)\Delta_C(a)$$

$$= m_A \left(\sum_{(a)} \lambda_A(\epsilon_C(a_{(1)})) \otimes \phi(a_{(2)}) \right)$$

$$= \sum_{(a)} \lambda_A(\epsilon_C(a_{(1)}))\phi(a_{(2)})$$

$$= \sum_{(a)} \epsilon_C(a_{(1)})\lambda_A(1_K)\phi(a_{(2)})$$

$$= \sum_{(a)} \epsilon_C(a_{(1)})1_A\phi(a_{(2)})$$

$$= \sum_{(a)} \phi(\epsilon_C(a_{(1)})a_{(2)})$$

$$= \phi(a) \quad \text{by the counit property.}$$

Thus, $\lambda_A \epsilon_C * \phi = \phi$. In a similar manner, one also obtains $\phi * \lambda_A \epsilon_C = \phi$, and so, $\mathrm{Hom}_K(C, A)$ is a monoid under $*$. $\square$

Proposition 3.1.7. Let H be a K-Hopf algebra and let $\mathrm{Hom}_K(H, H)$ be the monoid under convolution $*$. Then

$$\sigma_H * I_H = \lambda_H \epsilon_H = I_H * \sigma_H.$$

In other words, σ_H is a left and right inverse of I_H under $*$.

Proof. Exercise. $\square$

Convolution can be used to show that the coinverse map $\sigma_H : H \to H$ is an algebra "antihomomorphism" (and hence σ_H is an algebra homomorphism if H is commutative).

Proposition 3.1.8. *Let H be a K-Hopf algebra with coinverse σ_H. Then the following properties hold.*

(i) $\sigma_H(ab) = \sigma_H(b)\sigma_H(a)$ for all $a, b \in H$,
(ii) $\sigma_H(1_H) = 1_H$.

Proof. For (i), we consider the K-coalgebra $H \otimes H$ and the collection of linear transformations $\text{Hom}_K(H \otimes H, H)$. We have $m_H \in \text{Hom}_K(H \otimes H, H)$. We define two additional elements of $\text{Hom}_K(H \otimes H, H)$ as follows: For $a, b \in H$,

$$\sigma_H m_H : H \otimes H \to H, \quad a \otimes b \mapsto \sigma_H(ab),$$

$$\phi = m_H(\sigma_H \otimes \sigma_H)\tau : H \otimes H \to H, \quad a \otimes b \mapsto \sigma_H(b)\sigma_H(a).$$

Recall that $H \otimes H$ is a coalgebra with comultiplication

$$\Delta_{H \otimes H} = (I_H \otimes \tau \otimes I_H)(\Delta_H \otimes \Delta_H)$$

given as $a \otimes b \mapsto \sum_{(a,b)} a_{(1)} \otimes b_{(1)} \otimes a_{(2)} \otimes b_{(2)}$, and counit $\epsilon_{H \otimes H}$ defined by $a \otimes b \mapsto \epsilon_H(a)\epsilon_H(b)$.

For the proof of (i) we first show that

$$m_H * \phi = \lambda_H \epsilon_{H \otimes H} = \phi * m_H, \tag{3.2}$$

and

$$m_H * \sigma_H m_H = \lambda_H \epsilon_{H \otimes H} = \sigma_H m_H * m_H. \tag{3.3}$$

We first prove (3.2). For $a, b \in H$,

$$(m_H * \phi)(a \otimes b) = (m_H * m_H(\sigma_H \otimes \sigma_H)\tau))(a \otimes b)$$

$$= m_H(m_H \otimes m_H(\sigma_H \otimes \sigma_H)\tau))\Delta_{H \otimes H}(a \otimes b))$$

$$= m_H\left(\sum_{(a,b)} m_H(a_{(1)} \otimes b_{(1)}) \otimes m_H(\sigma_H \otimes \sigma_H)\tau(a_{(2)} \otimes b_{(2)})\right)$$

$$= \sum_{(a,b)} a_{(1)} b_{(1)} \sigma_H(b_{(2)})\sigma_H(a_{(2)})$$

$$= \sum_{(a)} a_{(1)} \epsilon_H(b)\sigma_H(a_{(2)}) \quad \text{by the coinverse property}$$

$$= \sum_{(a)} \epsilon_H(b) a_{(1)} \sigma_H(a_{(2)})$$

$$= \epsilon_H(b) \epsilon_H(a) 1_H \quad \text{by the coinverse property}$$

$$= \lambda_H \epsilon_{H \otimes H}(a \otimes b),$$

thus $m_H * \phi = \lambda_H \epsilon_{H \otimes H}$. A similar calculation yields $\phi * m_H = \lambda_H \epsilon_{H \otimes H}$.
We next consider (3.3). For $a, b \in H$,

$$(m_H * \sigma_H m_H)(a \otimes b) = m_H(m_H \otimes \sigma_H m_H) \Delta_{H \otimes H}(a \otimes b)$$

$$= m_H \left(\sum_{(a,b)} m_H(a_{(1)} \otimes b_{(1)}) \otimes \sigma_H m_H(a_{(2)} \otimes b_{(2)}) \right)$$

$$= \sum_{(a,b)} a_{(1)} b_{(1)} \sigma_H(a_{(2)} b_{(2)})$$

$$= m_H(I_H \otimes \sigma_H) \Delta_H(ab)$$

$$= \epsilon_H(ab) 1_H \quad \text{by the coinverse property}$$

$$= \epsilon_H(a) \epsilon_H(b) 1_H$$

$$= \lambda_H \epsilon_{H \otimes H},$$

thus $m_H * \sigma_H m_H = \lambda_H \epsilon_{H \otimes H}$. A similar computation shows that $\sigma_H m_H * m_H = \lambda_H \epsilon_{H \otimes H}$.
Now, from (3.2) and (3.3) we have

$$m_H * \phi = m_H * \sigma_H m_H,$$

thus

$$(\phi * m_H) * \phi = (\phi * m_H) * \sigma_H m_H$$

$$\lambda_H \epsilon_{H \otimes H} * \phi = \lambda_H \epsilon_{H \otimes H} * \sigma_H m_H$$

$$\phi = \sigma_H m_H \quad \text{by Proposition 3.1.6.}$$

Thus (i) is proved.
For (ii) observe that

$$1_H = 1_K 1_H$$

$$= \epsilon_H(1_H) 1_H$$

$$= m_H(I_H \otimes \sigma_H) \Delta_H(1_H)$$

$$= m_H(I_H \otimes \sigma_H)(1_H \otimes 1_H)$$

$$= \sigma_H(1_H).$$

$\square$

We can now show that the cocommutativity of H implies that σ_H has order 2.

Proposition 3.1.9. *Let H be a K-Hopf algebra with coinverse σ_H. If H is cocommutative, then $\sigma_H^2 = I_H$.*

Proof. Let $\text{Hom}_K(H, H)$ be the collection of linear transformations endowed with convolution $*$. Note that σ_H, σ_H^2, I_H, and $\lambda_H \epsilon_H$ are all elements of $\text{Hom}_K(H, H)$. For $a \in H$,

$$(\sigma_H * \sigma_H^2)(a) = \sum_{(a)} \sigma_H(a_{(2)})\sigma_H(\sigma_H(a_{(1)})), \quad \text{since } H \text{ is cocommutative}$$

$$= \sum_{(a)} \sigma_H(\sigma_H(a_{(1)})a_{(2)}), \quad \text{by Proposition 3.1.8(i)}$$

$$= \sigma_H(\epsilon_H(a)1_H) \quad \text{by the coinverse property}$$

$$= \epsilon_H(a)1_H \quad \text{by Proposition 3.1.8(ii)}$$

$$= (\lambda_H \epsilon_H)(a).$$

Thus

$$I_H * (\sigma_H * \sigma_H^2) = I_H * (\lambda_H \epsilon_H)$$

$$= I_H \quad \text{by Proposition 3.1.6.}$$

On the other hand,

$$I_H * (\sigma_H * \sigma_H^2) = (I_H * \sigma_H) * \sigma_H^2$$

$$= \lambda_H \epsilon_H * \sigma_H^2 \quad \text{by Proposition 3.1.7}$$

$$= \sigma_H^2 \quad \text{by Proposition 3.1.6,}$$

and so, $\sigma_H^2 = I_H$. $\square$

Let H be a K-Hopf algebra with coinverse σ_H. Then σ_H is a coalgebra "antihomomorphism" (in fact, σ_H is a K-coalgebra homomorphism if H is cocommutative).

Proposition 3.1.10. *Let H be a K-Hopf algebra with coinverse σ_H. Then the following properties hold.*

(i) $\tau(\sigma_H \otimes \sigma_H)\Delta_H = \Delta_H \sigma_H$,
(ii) $\epsilon_H \sigma_H = \epsilon_H$.

Proof. We consider the collection of linear transformations $\text{Hom}_K(H, H \otimes H)$. We note that $\Delta_H \in \text{Hom}_K(H, H \otimes H)$ and define two additional elements of $\text{Hom}_K(H, H \otimes H)$, they are:

$$\psi = \tau(\sigma_H \otimes \sigma_H)\Delta_H : H \to H \otimes H, \quad a \mapsto \sum_{(a)} \sigma_H(a_{(2)}) \otimes \sigma_H(a_{(1)}),$$

$$\Delta_H \sigma_H : H \to H \otimes H, \quad a \mapsto \sum_{(\sigma_H(a))} \sigma_H(a)_{(1)} \otimes \sigma_H(a)_{(2)},$$

for $a \in H$. Recall that $H \otimes H$ is a K-algebra with multiplication $m_{H \otimes H}$ defined as

$$(a \otimes b) \otimes (c \otimes d) \mapsto ac \otimes bd,$$

and unit $\lambda_{H \otimes H}$ given as $r \mapsto \lambda_H(r) \otimes 1_H = r1_H \otimes 1_H$, for $a, b, c, d \in H, r \in K$.

To prove (i) we first establish that

$$\Delta_H * \psi = \lambda_{H \otimes H} \epsilon_H = \psi * \Delta_H \tag{3.4}$$

and

$$\Delta_H * \Delta_H \sigma_H = \lambda_{H \otimes H} \epsilon_H = \Delta_H \sigma_H * \Delta_H. \tag{3.5}$$

For (3.4), for $a \in H$,

$$(\Delta_H * \psi)(a) = m_{H \otimes H}(\Delta_H \otimes \psi)\Delta_H(a)$$

$$= m_{H \otimes H}(\Delta_H \otimes \psi)\left(\sum_{(a)} a_{(1)} \otimes a_{(2)}\right)$$

$$= m_{H \otimes H}\left(\sum_{(a)} \Delta_H(a_{(1)}) \otimes \psi(a_{(2)})\right)$$

$$= m_{H \otimes H}\left(\sum_{(a)} a_{(1)} \otimes a_{(2)} \otimes \sigma_H(a_{(4)}) \otimes \sigma_H(a_{(3)})\right)$$

$$= \sum_{(a)} a_{(1)}\sigma_H(a_{(4)}) \otimes a_{(2)}\sigma_H(a_{(3)})$$

$$= \sum_{(a)} a_{(1)}\sigma_H(a_{(3)}) \otimes \epsilon_H(a_{(2)})1_H \quad \text{by the coinverse property}$$

$$= \sum_{(a)} \epsilon_H(a_{(2)})a_{(1)}\sigma_H(a_{(3)}) \otimes 1_H$$

$$= \sum_{(a)} a_{(1)}\sigma_H(a_{(2)}) \otimes 1_H \quad \text{by the counit property}$$

$$= \epsilon_H(a)1_H \otimes 1_H \quad \text{by the coinverse property}$$

$$= (\lambda_{H \otimes H}\epsilon_H)(a).$$

Thus, $\Delta_H * \psi = \lambda_{H \otimes H}\epsilon_H$. A similar computation shows that $\psi * \Delta_H = \lambda_{H \otimes H}\epsilon_H$, thus (3.4) holds. A straightforward computation also yields (3.5). It follows that

$$\psi * (\Delta_H * \psi) = \psi * (\Delta_H * \Delta_H \sigma_H)$$

$$(\psi * \Delta_H) * \psi = (\psi * \Delta_H) * \Delta_H \sigma_H$$

$$\lambda_{H \otimes H} \epsilon_H * \psi = \lambda_{H \otimes H} \epsilon_H * \Delta_H \sigma_H$$

$$\psi = \Delta_H \sigma_H,$$

which yields (i).

For (ii): For all $a \in H$,

$$\epsilon_H(a) = \epsilon_H(a) \epsilon_H(1_H)$$

$$= \epsilon_H(\epsilon_H(a) 1_H)$$

$$= \epsilon_H \left(\sum_{(a)} a_{(1)} \sigma_H(a_{(2)}) \right) \quad \text{by the coinverse property}$$

$$= \sum_{(a)} \epsilon_H(a_{(1)}) \epsilon_H(\sigma_H(a_{(2)}))$$

$$= \sum_{(a)} \epsilon_H(\sigma_H(\epsilon_H(a_{(1)}) a_{(2)}))$$

$$= \epsilon_H(\sigma_H(a)) \quad \text{by the counit property.}$$

$$\square$$

The coalgebra anti-homomorphism property of σ_H can be used to show that $\sigma_H^2 = I_H$ whenever H is commutative, see §3.5, Exercise 6. We will need the coalgebra anti-homomorphism property of σ_H in §3.2.

Let H be a K-Hopf algebra. A **Hopf ideal** I is a biideal (viewing H as a K-bialgebra) that satisfies $\sigma_H(I) \subseteq I$.

Proposition 3.1.11. *Let $I \subseteq H$ be a Hopf ideal of H. Then H/I is a K-Hopf algebra.*

Proof. By Proposition 2.1.5, H/I is a K-bialgebra. Since $\sigma_H(I) \subseteq I$, σ_H induces a K-linear map $\sigma_{H/I} : H/I \to H/I$ defined as $a + I \mapsto \sigma_H(a) + I$. One has

$$m_{H/I}(I_{H/I} \otimes \sigma_{H/I}) \Delta_{H/I}(a + I) = m_{H/I}(I_{H/I} \otimes \sigma_{H/I}) \left(\sum_{(a)} a_{(1)} + I \otimes a_{(2)} + I \right)$$

$$= \sum_{(a)} (a_{(1)} \sigma_H(a_{(2)})) + I$$

$$= \epsilon_H(a) 1_H + I$$

$$= \epsilon_{H/I}(a + I) 1_{H/I}.$$

In a similar manner, one obtains

$$m_{H/I}(\sigma_{H/I} \otimes I_{H/I}) \Delta_{H/I}(a + I) = \epsilon_{H/I}(a + I) 1_{H/I}.$$

Thus H/I is a K-Hopf algebra. $\square$

Let H and H' be K-Hopf algebras. A bialgebra homomorphism $\phi : H \to H'$ is a **homomorphism of Hopf algebras** if

$$\phi(\sigma_H(a)) = \sigma_{H'}(\phi(a))$$

for all $a \in H$. The Hopf homomorphism ϕ is an **isomorphism of Hopf algebras** if ϕ is a bijection.

Proposition 3.1.12. *Let H be a finite dimensional vector space over the field K. Then H is a K-Hopf algebra if and only if H^* is a K-Hopf algebra.*

Proof. Suppose that H is a K-Hopf algebra. By Proposition 2.1.10, H^* is a bialgebra. Let $\sigma_H^* : H^* \to H^*$ be the transpose of the coinverse map σ_H. Then for $f \in H^*, a \in H$,

$$
\begin{aligned}
(m_{H^*}(I_{H^*} \otimes \sigma_H^*)\Delta_{H^*}(f))(a) &= m_{H^*}((I_{H^*} \otimes \sigma_H^*)\Delta_{H^*}(f))(a) \\
&= ((I_{H^*} \otimes \sigma_H^*)\Delta_{H^*}(f))\Delta_H(a) \\
&= \Delta_{H^*}(f)(I_H \otimes \sigma_H)\Delta_H(a) \\
&= f(m_H(I_H \otimes \sigma_H)\Delta_H(a)) \\
&= f(\epsilon_H(a)1_H) \quad \text{by the counit property.} \\
&= \epsilon_H(a)f(1_H) \\
&= \epsilon_{H^*}(f)\epsilon_H(a) \\
&= \epsilon_{H^*}(f)1_{H^*}(a).
\end{aligned}
$$

In a similar manner one obtains

$$m_{H^*}(\sigma_H^* \otimes I_{H^*})\Delta_{H^*}(f) = \epsilon_{H^*}(f)1_{H^*}.$$

Consequently, $\sigma_{H^*} = \sigma_H^*$ satisfies the coinverse property and H^* is a K-Hopf algebra.

Conversely, if H^* is a K-Hopf algebra, then $H^{**} = H$ is a K-Hopf algebra. $\square$

Example 3.1.13. Let G be a finite group. Then KG is a K-Hopf algebra (Example 3.1.2). Thus KG^* is a K-Hopf algebra. The subset $\{p_v\}_{v \in G}$ with $p_v(\tau) = \delta_{v,\tau}$ is a K-basis for KG^*. The multiplication on KG^* is defined as $(p_v p_\tau)(\omega) = p_v(\omega)p_\tau(\omega)$, hence $p_v p_\tau = \delta_{v,\tau}p_v$. The unit map of KG^*, $\lambda_{KG^*} : K \to KG^*$ is defined by $r \mapsto r\epsilon_{KG}$. The comultiplication of KG^* is given as

$$\Delta_{KG^*}(p_v) = \sum_{v=\tau\omega} p_\tau \otimes p_\omega,$$

cf. Proposition 1.3.10, and the counit map $\epsilon_{KG^*} : KG^* \to K$ is defined as $p_v \mapsto p_v(1)$. Finally, the coinverse map $\sigma_{KG^*} : KG^* \to KG^*$ is given as $p_v \mapsto p_{v^{-1}}$.

Let p be a prime number, let $n \geq 1$ be an integer. Let K be a field containing ζ_{p^n}, a primitive p^nth root of unity. Let $G = C_{p^n}$ denote the cyclic group of order p^n generated by g. Then the linear dual $KC_{p^n}^*$ is a K-Hopf algebra as in Example 3.1.13.

Proposition 3.1.14. $KC_{p^n}^* \cong KC_{p^n}$, as K-Hopf algebras.

Proof. Let $\{p_i\}_{i=0}^{p^n-1}$ be the basis for $KC_{p^n}^*$ dual to the basis $\{g^i\}_{i=0}^{p^n-1}$ for KC_{p^n}. Let $\phi : KC_{p^n}^* \to KC_{p^n}$ be the K-linear map defined by $p_i = \frac{1}{p^n} \sum_{j=0}^{p^n-1} \zeta_{p^n}^{-ij} g^j$. Then as the reader can show, ϕ is an isomorphism of K-Hopf algebras, see §3.5, Exercise 14. $\square$

3.2   Integrals and Hopf Modules

In this section we define the set of left and right integrals of a Hopf algebra over K. We show that the set of left (or right) integrals $\int_H^l$ (or $\int_H^r$) is a subspace of H and an ideal of H. We compute the ideal of integrals for KG and KG^* in the case that G is a finite group.

Next we specialize to the case where H is finite dimensional and define the concept of a right comodule over H, a vector space that is "opposite" to a right H-module. We show that H^* is a right H-comodule. We then consider right Hopf modules over H, which are vector spaces that are both right H-modules and right H-comodules. We state the Fundamental Theorem of Hopf Modules, which says that every right Hopf module over a finite dimensional Hopf algebra H is isomorphic to a trivial right Hopf module of the form $W \otimes H$ for some vector space W. We show that the right H-comodule H^* is a right Hopf module and prove a special case of the Fundamental Theorem which says that $H^* \cong W \otimes H$, where $W = \int_{H^*}^l$ and $\int_{H^*}^l \otimes H$ has a trivial right Hopf module structure. We conclude that $\int_{H^*}^l$ is a one-dimensional subspace of H^*, and so H^* admits a generating integral.

$$* \quad * \quad *$$

Let H be an K-Hopf algebra.

Definition 3.2.1.   A **left integral** of H is an element $y \in H$ that satisfies

$$xy = \epsilon(x)y,$$

for all $x \in H$. A **right integral** of H is an element $y \in H$ that satisfies

$$yx = \epsilon(x)y,$$

for all $x \in H$.

We denote the collection of left integrals of H by $\int_H^l$ and the collection of right integrals of H by $\int_H^r$.

Proposition 3.2.2. *The set of left integrals $\int_H^l$ is a subspace of H; the set of right integrals $\int_H^r$ is a subspace of H.*

Proof. We prove the result for left integrals. We first show that $\int_H^l$ is an additive subgroup of H. Let $x, y \in \int_H^l$. Then for $z \in H$, we have

$$z(x + y) = zx + zy = \epsilon(z)x + \epsilon(z)y = \epsilon(z)(x + y),$$

and so $\int_H^l$ is closed under addition. Moreover, $0 \in \int_H^l$ since $z0 = 0 = \epsilon(z)0$, and $-x \in \int_H^l$ since $z(-x) = -(zx) = -(\epsilon(z)x) = \epsilon(z)(-x)$. Thus $\int_H^l \leq H$. Now for $r \in K, x \in \int_H^l$,

$$z(rx) = (rz)x = \epsilon_H(rz)x = r\epsilon_H(x)x = \epsilon_H(z)(rx),$$

and so, $\int_H^l$ is a K-subspace of H. The result for $\int_H^r$ is proved in the same way and we leave the details to the reader.   □

Proposition 3.2.3. $\int_H^l$ *is an ideal of H; $\int_H^r$ is an ideal of H.*

Proof. We prove the result for $\int_H^l$. By Proposition 3.2.2 $\int_H^l$ is an additive subgroup of H. We show that $H\int_H^l \subseteq \int_H^l$ and $\int_H^l H \subseteq \int_H^l$. Let $x \in H$, $y \in \int_H^l$. Then for $z \in H$,

$$z(xy) = (zx)y = \epsilon_H(zx)y = \epsilon_H(z)\epsilon_H(x)y = \epsilon_H(z)(xy),$$

and thus, $xy \in \int_H^l$. Moreover,

$$z(yx) = (zy)x = (\epsilon_H(z)y)x = \epsilon_H(z)(yx),$$

and thus $yx \in \int_H^l$. The result for $\int_H^r$ is proved in the same way and we leave the details to the reader.   □

A K-Hopf algebra H is **unimodular** if $\int_H^l = \int_H^r$. If H is commutative, then H is unimodular. Commutativity is not a necessary condition for H to be unimodular, however.

Proposition 3.2.4. *Let G be a finite group and let KG be the group ring of G over K. Then $\int_{KG}^l = \int_{KG}^r = K\sum_{\tau \in G}\tau$.*

Proof. Let $L = K\sum_{\tau \in G}\tau$ and let $a\sum_{\tau \in G}\tau \in L$, $a \in K$. Since $\{v\}_{v \in G}$ is a K-basis for KG, an element $x \in KG$ can be written as $x = \sum_{v \in G}x_v v$, for $x_v \in K$. Now

$$\left(\sum_{v \in G}x_v v\right)\left(a\sum_{\tau \in G}\tau\right) = \sum_{v \in G}\left(x_v a v\sum_{\tau \in G}\tau\right)$$

$$= \left(\sum_{v \in G}x_v\right)\left(a\sum_{\tau \in G}\tau\right)$$

$$= \epsilon_{KG}\left(\sum_{v \in G}x_v v\right)\left(a\sum_{\tau \in G}\tau\right),$$

and so, $L \subseteq \int_{KG}^l$.

Now suppose $x = \sum_{\tau \in G} x_\tau \tau \in \int_{KG}^l$. For $\upsilon \in G$,

$$x = \epsilon_{KG}(\upsilon)x = \upsilon x = \sum_{\tau \in G} x_\tau \upsilon \tau = \sum_{\tau \in G} x_\tau \rho(\tau)$$

where $\rho : G \to G$ is a permutation of the elements of G. Thus, $x_\tau = a$ for some $a \in K$ and all $\tau \in G$, and so, $\int_{KG}^l \subseteq L$.

An analogous argument is used to show that $\int_{KG}^r = L$. □

Since G is finite, the linear dual KG^* is a K-Hopf algebra (Proposition 3.1.12), thus KG^* is unimodular since KG^* is commutative. We compute $\int_{KG^*}^l = \int_{KG^*}^r$ below.

Proposition 3.2.5. $\int_{KG^*}^l = \int_{KG^*}^r = Kp_1$, where p_1 is the element of KG^* defined by $p_1(\upsilon) = 1$ if $\upsilon = 1$, $p_1(\upsilon) = 0$ if $\upsilon \neq 1$.

Proof. Recall that $\{p_\upsilon : \upsilon \in G\}$ defined by $p_\upsilon(\tau) = \delta_{\upsilon,\tau}$ is a K-basis for KG^*. Let $ap_1 \in Kp_1$ and let $\sum_{\upsilon \in G} x_\upsilon p_\upsilon \in KG^*$. Now

$$\left(\sum_{\upsilon \in G} x_\upsilon p_\upsilon\right) ap_1 = ax_1 p_1 = x_1 ap_1 = \epsilon_{KG^*}\left(\sum_{\upsilon \in G} x_\upsilon p_\upsilon\right) ap_1,$$

and so, $Kp_1 \subseteq \int_{KG^*}^l$.

Next, assume that $\sum_{\tau \in G} x_\tau p_\tau \in \int_{KG^*}^l$. Then for $\upsilon \in G$,

$$p_\upsilon\left(\sum_{\tau \in G} x_\tau p_\tau\right) = x_\upsilon p_\upsilon = \epsilon_{KG^*}(p_\upsilon) \sum_{\tau \in G} x_\tau p_\tau = \delta_{\upsilon,1} \sum_{\tau \in G} x_\tau p_\tau,$$

and so $x_\tau = 0$ for all $\tau \neq 1$. Hence $\int_{KG^*}^l \subseteq Kp_1$, and so, $\int_{KG^*}^l = Kp_1$. A similar argument shows that $\int_{KG^*}^r = Kp_1$. □

For the remainder of this section, all Hopf algebras will be finite dimensional over a field K.

Definition 3.2.6. Let H be a K-Hopf algebra and let M be a vector space over K. Then M is a **right H-comodule** if there exists a K-linear map $\Psi : M \to M \otimes H$ for which

(i) $(\Psi \otimes I_H)\Psi = (I_M \otimes \Delta_H)\Psi$,
(ii) $(I_M \otimes \epsilon_H)\Psi(m) = m \otimes 1_K$, $\forall m \in M$.

Example 3.2.7. The K-Hopf algebra H is a right comodule over itself with the comultiplication map Δ_H playing the role of Ψ.

Let M, N be right H-comodules with comodule maps Ψ_M, Ψ_N, respectively. A **homomorphism of right comodules** is a linear transformation $\phi : M \to N$ for which

$$\Psi_N \circ \phi = (\phi \otimes I_H) \circ \Psi_M.$$

We adapt Sweedler notation for comodules. Let M be a right H-comodule with structure map $\Psi : M \rightarrow M \otimes H$. Extending Sweedler notation for comultiplication, we write

$$\Psi(\beta) = \sum_{(\beta)} \beta_{(1)} \otimes b_{(2)}$$

for $\beta, \beta_{(1)} \in M$, $b_{(2)} \in H$. We shall write

$$\Psi(\beta_{(1)}) = \sum_{(\beta_{(1)})} \beta_{(1)(1)} \otimes b_{(1)(2)}.$$

Now,

$$(\Psi \otimes I_H)\Psi(\beta) = (\Psi \otimes I_H)\left(\sum_{(\beta)} \beta_{(1)} \otimes b_{(2)} \right)$$

$$= \sum_{(\beta, \beta_{(1)})} \beta_{(1)(1)} \otimes b_{(1)(2)} \otimes b_{(2)}, \qquad (3.6)$$

and

$$(I_M \otimes \Delta_H)\Psi(\beta) = (I_M \otimes \Delta_H)\left(\sum_{(\beta)} \beta_{(1)} \otimes b_{(2)} \right)$$

$$= \sum_{(\beta, b_{(2)})} \beta_{(1)} \otimes b_{(2)(1)} \otimes \beta_{(2)(2)}. \qquad (3.7)$$

By Definition 3.2.6(i) the expressions in (3.6) and (3.7) are equal. The common value in (3.6) and (3.7) will be denoted as

$$\sum_{(\beta)} \beta_{(1)} \otimes b_{(2)} \otimes b_{(3)}.$$

Similarly, the common value of

$$(I_M \otimes I_H \otimes \Delta)(I_M \otimes \Delta)\Psi(\beta) = (I_M \otimes \Delta \otimes I_H)(I_M \otimes \Delta)\Psi(\beta)$$

$$= (\Psi \otimes I_H \otimes I_H)(I_M \otimes \Delta)\Psi(\beta)$$

$$= (I_M \otimes I_H \otimes \Delta)(\Psi \otimes I_H)\Psi(\beta)$$

$$= (I_M \otimes \Delta \otimes I_H)(\Psi \otimes I_H)\Psi(\beta)$$

$$= (\Psi \otimes I_H \otimes I_H)(\Psi \otimes I_H)\Psi(\beta)$$

is denoted as

$$\sum_{(\beta)} \beta_{(1)} \otimes b_{(2)} \otimes b_{(3)} \otimes b_{(4)}.$$

We have seen that H is a right comodule over itself. Here is another example of a right comodule.

Example 3.2.8. Let G be a finite group and let KG be the K-Hopf algebra of Example 3.1.2, and let KG^* be the K-Hopf algebra of Example 3.1.13. Then $\{p_\upsilon\}_{\upsilon \in G}$ is the basis for KG^* dual to the basis $\{\upsilon\}_{\upsilon \in G}$ for KG. Let $\Psi : KG^* \to KG^* \otimes KG$ be the K-linear map defined by

$$\Psi(p_\upsilon) = \sum_{\tau \in G} p_\tau p_\upsilon \otimes \tau$$

$$= p_\upsilon \otimes \upsilon.$$

Then KG^* is a right KG-comodule with structure map Ψ.

We can generalize Example 3.2.8 to finite dimensional Hopf algebras. Let $\{b_i\}_{i=1}^n$ be a basis for H and let $\{\alpha_i\}_{i=1}^n$ be the basis for H^* dual to the basis $\{b_i\}_{i=1}^n$. We have $\alpha_i(b_j) = b_j(\alpha_i) = \delta_{i,j}$ (note: $b_j \in H^{**} = H$) and $\beta = \sum_{i=1}^n b_i(\beta)\alpha_i$ for $\beta \in H^*$.

Proposition 3.2.9. H^* *is a right H-comodule with structure map*

$$\Psi : H^* \to H^* \otimes H$$

defined as

$$\Psi(\beta) = \sum_{i=1}^n \alpha_i \beta \otimes b_i,$$

for $\beta \in H^*$.

Proof. We show that conditions (i) and (ii) of Definition 3.2.6 hold. For (i): first note that an element $\sum \gamma \otimes x \otimes y \in H^* \otimes H \otimes H$ defines a K-linear map

$$H^* \otimes H^* \to H^*$$

by the rule

$$\left(\sum \gamma \otimes x \otimes y\right)(\tau \otimes \rho) = x(\tau)y(\rho)\gamma,$$

for $\tau, \rho \in H^*$. Let $\beta \in H^*$ and note that

$$(I_{H^*} \otimes \Delta_H)\Psi(\beta) = \sum_{i=1}^n \alpha_i \beta \otimes \Delta_H(b_i)$$

$$= \sum_{i=1}^n \sum_{(b_i)} \alpha_i \beta \otimes b_{i(1)} \otimes b_{i(2)}$$

is an element of $H^* \otimes H \otimes H$. Its value at $\tau \otimes \rho$ is

$$((I_{H^*} \otimes \Delta_H)\Psi(\beta))(\tau \otimes \rho) = \left(\sum_{i=1}^{n} \sum_{(b_i)} \alpha_i \beta \otimes b_{i(1)} \otimes b_{i(2)} \right)(\tau \otimes \rho)$$

$$= \sum_{i=1}^{n} \sum_{(b_i)} b_{i(1)}(\tau)b_{i(2)}(\rho)\alpha_i \beta$$

$$= \sum_{i=1}^{n} b_i(\tau\rho)\alpha_i \beta$$

$$= \tau\rho\beta.$$

On the other hand, $(\Psi \otimes I_H)\Psi(\beta) \in H^* \otimes H \otimes H$ and its value at $\tau \otimes \rho$ is

$$((\Psi \otimes I_H)\Psi(\beta))(\tau \otimes \rho) = \left(\sum_{i=1}^{n} \Psi(\alpha_i \beta) \otimes b_i \right)(\tau \otimes \rho)$$

$$= \left(\sum_{i=1}^{n} \sum_{j=1}^{n} \alpha_j(\alpha_i\beta) \otimes b_j \otimes b_i \right)(\tau \otimes \rho)$$

$$= \sum_{i=1}^{n} \sum_{j=1}^{n} b_j(\tau)b_i(\rho)\alpha_j\alpha_i\beta$$

$$= \left(\sum_{j=1}^{n} b_j(\tau)\alpha_j \right)\left(\sum_{i=1}^{n} b_i(\rho)\alpha_i \right)\beta$$

$$= \tau\rho\beta.$$

which proves condition (i) of Definition 3.2.6. For condition (ii) of Definition 3.2.6:

$$(I_{H^*} \otimes \epsilon_H)\Psi(\beta) = \sum_{i=1}^{n} \alpha_i\beta \otimes \epsilon_H(b_i)$$

$$= \sum_{i=1}^{n} \epsilon_H(b_i)\alpha_i\beta \otimes 1_K$$

$$= \left(\sum_{i=1}^{n} \epsilon_H(b_i)\alpha_i\beta \right) \otimes 1_K$$

$$= \left(\sum_{i=1}^{n} b_i(\epsilon_H)\alpha_i\beta \right) \otimes 1_K$$

$$= \epsilon_H\beta \otimes 1_K$$

$$= \beta \otimes 1_K,$$

since $\epsilon_H = 1_{H^*}$. $\qquad\square$

The right H-comodule structure on H^* given in Proposition 3.2.9 induces a left H^*-module structure on H^* defined as

$$\alpha \cdot \beta = s_2(I_{H^*} \otimes \alpha)\Psi(\beta) = \sum_{i=1}^n \alpha(b_i)\alpha_i\beta,$$

for $\beta, \alpha \in H^*$. Here, s_2 is the map $H^* \otimes K \to H^*$ given as $\gamma \otimes r \mapsto r\gamma$, for $\gamma \in H^*$, $r \in K$. The induced left H^*-module structure is precisely the left multiplication action of H^* on itself since

$$\alpha \cdot \beta = \sum_{i=1}^n \alpha(b_i)\alpha_i\beta = \left(\sum_{i=1}^n b_i(\alpha)\alpha_i \right)\beta = \alpha\beta. \tag{3.8}$$

Let $\Psi : H^* \to H^* \otimes H$ be the right H-comodule structure map of Proposition 3.2.9. In Sweedler notation, formula (3.8) is now written as

$$\alpha\beta = \sum_{(\beta)} \alpha(b_{(2)})\beta_{(1)}. \tag{3.9}$$

Let M be a right H-comodule with structure map $\Psi : M \to M \otimes H$. If Ψ respects certain H-module structures on M and $M \otimes H$, then M is a "right Hopf module" over H. Here is a precise definition.

Definition 3.2.10. Let H be an K-Hopf algebra and let M be a vector space over K. Then M is a **right Hopf module over H** if

  (i) M is a right H-module with scalar multiplication given as "$\cdot$,"
  (ii) M is a right H-comodule with structure map $\Psi : M \to M \otimes H$,
 (iii) the right H-comodule structure map $\Psi : M \to M \otimes H$ is a homomorphism of right H-modules where $M \otimes H$ is a right H-module with scalar multiplication

$$(m \otimes k)h = \sum_{(h)} m \cdot h_{(1)} \otimes kh_{(2)},$$

for $h, k \in H, m \in M$.

Example 3.2.11. Let H be a K-Hopf algebra. Then H is a right module over itself through right multiplication and a right comodule over itself with structure map Δ_H. Endow $H \otimes H$ with the structure of a right H-module with scalar multiplication given as

$$(a \otimes b)h = \sum_{(h)} ah_{(1)} \otimes bh_{(2)},$$

for $a, b, h \in H$. Then $\Delta_H : H \to H \otimes H$ is a homomorphism of right H-modules (§3.5, Exercise 17). Thus H is a right Hopf module over itself.

Example 3.2.12. Let W be a vector space over K and a right H-module with action denoted as ".". Let $M = W \otimes H$, and on $W \otimes H$ define a right H-module structure by

$$(w \otimes a)h = \sum_{(h)} w \cdot h_{(1)} \otimes ah_{(2)},$$

for $w \in W$, $a, h \in H$. By §3.5, Exercise 18, $W \otimes H$ is a right H-comodule with structure map

$$I_W \otimes \Delta_H : W \otimes H \to W \otimes H \otimes H.$$

Endow $W \otimes H \otimes H$ with a right H-module structure defined as

$$(w \otimes a \otimes b)h = \sum_{(h)} (w \otimes a)h_{(1)} \otimes bh_{(2)}$$

$$= \sum_{(h, h_{(1)})} w \cdot h_{(1)} \otimes ah_{(2)} \otimes bh_{(3)},$$

for $w \in W$, $a, b, h \in H$. Now $I_W \otimes \Delta_H$ is an H-module homomorphism and so $W \otimes H$ is a right Hopf module over H.

Example 3.2.13 (Trivial Right Hopf Module). In Example 3.2.12 we take a vector space W with right H-module action defined as

$$w \cdot h = \epsilon_H(h)w,$$

for $w \in W$, $h \in H$; W is a **trivial right module**. Now the right action of H on $W \otimes H$ is

$$(w \otimes a)h = \sum_{(h)} w \cdot h_{(1)} \otimes ah_{(2)}$$

$$= \sum_{(h)} \epsilon_H(h_{(1)})w \otimes ah_{(2)}$$

$$= \sum_{(h)} w \otimes a\epsilon_H(h_{(1)})h_{(2)}$$

$$= w \otimes ah,$$

for $w \in W$, $a, h \in H$. Likewise, the right H-action on $W \otimes H \otimes H$ is

$$(w \otimes a \otimes b)h = \sum_{(h)} w \otimes ah_{(1)} \otimes bh_{(2)}.$$

The resulting right H-Hopf module $W \otimes H$ is a **trivial right Hopf module**.

Let M, N be right Hopf modules over H. Then the map $\phi : M \to N$ is a **homomorphism of Hopf modules** if ϕ is both a homomorphism of H-modules and a homomorphism of H-comodules.

Proposition 3.2.14 (Fundamental Theorem of Hopf Modules). *Let H be a K-Hopf algebra, let M be a right Hopf module over H with structure map $\Psi : M \to M \otimes H$. Let*

$$W = \{m \in M : \Psi(m) = m \otimes 1\},$$

and endow $W \otimes H$ with the structure of a trivial right Hopf module over H. Then $M \cong W \otimes H$ as right Hopf modules over H.

Proof. For the proof, the reader is referred to [Mo93, §1.9]. $\square$

Our goal is to show that the Fundamental Theorem of Hopf Modules holds in the case $M = H^*$; we show that there is an isomorphism of Hopf modules $H^* \cong W \otimes H$ where $W \otimes H$ is the trivial Hopf module with $W = \{\beta \in H^* : \Psi(\beta) = \beta \otimes 1\}$.

Of course, we first need to show that H^* is a right Hopf module over H. Put $\sigma = \sigma_H$. We define a right H-module structure on H^* by

$$(\beta \cdot h)(k) = \beta(k\sigma(h)) = \sum_{(\beta)} \beta_{(1)}(k)\beta_{(2)}(\sigma(h)) \tag{3.10}$$

and define a right H-module structure on $H^* \otimes H$ by

$$(\beta \otimes k)h = \sum_{(h)} \beta \cdot h_{(1)} \otimes kh_{(2)} \tag{3.11}$$

for $h, k \in H$, $\beta \in H^*$. The right action of H on H^* endows H^* with the structure of an "anti" right H-module algebra.

Lemma 3.2.15. *Let $h \in H$, $\alpha, \beta \in H^*$. Then*

$$(\alpha\beta) \cdot h = \sum_{(h)} (\alpha \cdot h_{(2)})(\beta \cdot h_{(1)}).$$

Proof. Since the comultiplication Δ_{H^*} is an R-algebra homomorphism,

$$((\alpha\beta) \cdot h)(k) = (\alpha\beta)(k\sigma(h))$$

$$= \sum_{(\alpha,\beta)} (\alpha_{(1)}\beta_{(1)})(k)(\alpha_{(2)}\beta_{(2)})(\sigma(h)),$$

for all $k \in H$. By Proposition 3.1.10(i),

$$\Delta(\sigma(h)) = \sum_{(h)} \sigma(h_{(2)}) \otimes \sigma(h_{(1)}),$$

thus

$$\sum_{(\alpha,\beta)} (\alpha_{(1)}\beta_{(1)})(k)(\alpha_{(2)}\beta_{(2)})(\sigma(h)) = \sum_{(\alpha,\beta,h)} (\alpha_{(1)}\beta_{(1)})(k)\alpha_{(2)}(\sigma(h_{(2)}))\beta_{(2)}(\sigma(h_{(1)}))$$

$$= \sum_{(\alpha,\beta,h,k)} \alpha_{(1)}(k_{(1)})\beta_{(1)}(k_{(2)})\alpha_{(2)}(\sigma(h_{(2)}))\beta_{(2)}(\sigma(h_{(1)}))$$

$$= \sum_{(\alpha,\beta,h,k)} \alpha_{(1)}(k_{(1)})\alpha_{(2)}(\sigma(h_{(2)}))\beta_{(1)}(k_{(2)})\beta_{(2)}(\sigma(h_{(1)}))$$

$$= \sum_{(h,k)} \alpha(k_{(1)}\sigma(h_{(2)}))\beta(k_{(2)}\sigma(h_{(1)}))$$

$$= \sum_{(h,k)} (\alpha \cdot h_{(2)})(k_{(1)})(\beta \cdot h_{(1)})(k_{(2)})$$

$$= \left(\sum_{(h)} (\alpha \cdot h_{(2)})(\beta \cdot h_{(1)})\right)(k)$$

which proves the lemma. $\qquad\square$

Proposition 3.2.16 (Larson and Sweedler, [LS69]). *Let H be a finite dimensional Hopf algebra with linear dual H^*. Let $\{b_i\}_{i=1}^n$ be a basis for H and let $\{\alpha_i\}_{i=1}^n$ be the basis for H^* dual to the basis $\{b_i\}_{i=1}^n$. Let H^* be a right H-module through (3.10), and let $H^* \otimes H$ be a right H-module through (3.11). Then H^* is a right Hopf module over H.*

Proof. By Proposition 3.2.9, H^* is a right H-comodule with structure map $\Psi :$ $H^* \to H^* \otimes H$ defined by $\Psi(\beta) = \sum_{i=1}^n \alpha_i\beta \otimes b_i$. Thus we only need to show that $\Psi(\beta \cdot h) = \Psi(\beta)h$, for $h \in H$, $\beta \in H^*$. For $\alpha \in H^*$ one has

$$\alpha(\beta \cdot h) = \alpha\left(\beta \cdot \sum_{(h)} \epsilon(h_{(2)})h_{(1)}\right), \quad \text{by the counit property}$$

$$= \sum_{(h)} (\alpha \cdot \epsilon(h_{(2)})1_H)(\beta \cdot h_{(1)})$$

$$= \sum_{(h)} (\alpha \cdot \sigma(h_{(2)})h_{(3)})(\beta \cdot h_{(1)}) \quad \text{by the coinverse property.}$$

Now, for $k \in H$,
$$\sum_{(h)} (\alpha \cdot \sigma(h_{(2)})h_{(3)})(\beta \cdot h_{(1)})(k)$$

$$= \sum_{(h,k)} (\alpha \cdot \sigma(h_{(2)})h_{(3)})(k_{(1)})(\beta \cdot h_{(1)})(k_{(2)})$$

$$= \sum_{(h,k)} \alpha(k_{(1)}\sigma(h_{(3)})\sigma(\sigma(h_{(2)})))(\beta \cdot h_{(1)})(k_{(2)}) \quad \text{by Proposition 3.1.8(i)}$$

$$= \sum_{(h,k)} \alpha(k_{(1)}\epsilon(\sigma(h_{(2)}))1_H)(\beta \cdot h_{(1)})(k_{(2)}) \quad \text{by the coinverse property}$$

$$= \sum_{(h,k)} \alpha(k_{(1)}\epsilon(h_{(2)})1_H)(\beta \cdot h_{(1)})(k_{(2)}) \quad \text{by Proposition 3.1.10(ii).}$$

Continuing with the calculation, we have
$\sum_{(h,k)} \alpha(k_{(1)}\epsilon(h_{(2)})1_H)(\beta \cdot h_{(1)})(k_{(2)})$

$$= \sum_{(h,k)} \alpha(k_{(1)}\sigma(h_{(2)})h_{(3)})(\beta \cdot h_{(1)})(k_{(2)}) \quad \text{by the coinverse property}$$

$$= \sum_{(h,k,\alpha)} \alpha_{(1)}(k_{(1)}\sigma(h_{(2)}))\alpha_{(2)}(h_{(3)})(\beta \cdot h_{(1)})(k_{(2)})$$

$$= \sum_{(h,k,\alpha)} \alpha_{(2)}(h_{(3)})(\alpha_{(1)} \cdot h_{(2)})(k_{(1)})(\beta \cdot h_{(1)})(k_{(2)})$$

$$= \sum_{(h,\alpha)} \alpha_{(2)}(h_{(3)})(\alpha_{(1)} \cdot h_{(2)})(\beta \cdot h_{(1)})(k)$$

Thus,

$$\alpha(\beta \cdot h) = \sum_{(h,\alpha)} \alpha_{(2)}(h_{(3)})(\alpha_{(1)} \cdot h_{(2)})(\beta \cdot h_{(1)}).$$

Now,

$$\alpha(\beta \cdot h) = \sum_{(h,\alpha)} \alpha_{(2)}(h_{(3)})(\alpha_{(1)} \cdot h_{(2)})(\beta \cdot h_{(1)})$$

$$= \sum_{(h,\alpha)} \alpha_{(2)}(h_{(2)})(\alpha_{(1)}\beta \cdot h_{(1)}) \quad \text{by Lemma 3.2.15}$$

$$= \sum_{(h,\alpha)} \alpha_{(2)}(h_{(2)})\left(\sum_{j=1}^{n} \alpha_{(1)}(b_j)\alpha_j\beta \cdot h_{(1)} \right) \quad \text{by (3.8)}$$

$$= \sum_{(h,\alpha)} \sum_{j=1}^{n} \alpha_{(1)}(b_j)\alpha_{(2)}(h_{(2)})(\alpha_j\beta \cdot h_{(1)})$$

$$= \sum_{(h)} \sum_{j=1}^{n} \alpha(b_jh_{(2)})(\alpha_j\beta \cdot h_{(1)}).$$

It follows that

$$\alpha_i(\beta \cdot h) \otimes b_i = \sum_{(h)} \sum_{j=1}^n \alpha_i(b_j h_{(2)})(\alpha_j \beta \cdot h_{(1)}) \otimes b_i$$

$$= \sum_{(h)} \sum_{j=1}^n (\alpha_j \beta \cdot h_{(1)}) \otimes \alpha_i(b_j h_{(2)}) b_i. \qquad (3.12)$$

Finally, we can show that Ψ respects H-module structure:

$$\Psi(\beta \cdot h) = \sum_{i=1}^n \alpha_i(\beta \cdot h) \otimes b_i$$

$$= \sum_{i=1}^n \sum_{(h)} \sum_{j=1}^n (\alpha_j \beta \cdot h_{(1)}) \otimes \alpha_i(b_j h_{(2)}) b_i \quad \text{by (3.12)}$$

$$= \sum_{(h)} \sum_{j=1}^n \sum_{i=1}^n (\alpha_j \beta \cdot h_{(1)}) \otimes \alpha_i(b_j h_{(2)}) b_i$$

$$= \sum_{(h)} \sum_{j=1}^n \alpha_j \beta \cdot h_{(1)} \otimes b_j h_{(2)}$$

$$= \left(\sum_{j=1}^n \alpha_j \beta \otimes b_j \right) h$$

$$= \Psi(\beta) h.$$

We conclude that H^* is a right Hopf module over H. $\qquad \square$

Example 3.2.17. Let G be a finite group and let KG be the K-Hopf algebra of Example 3.1.2. Then KG^* is a K-Hopf algebra as in Example 3.1.13. In this example, we illustrate Proposition 3.2.16 by showing that KG^* is a right Hopf module over KG. Now, KG^* is a right KG-module through the action

$$(p_\upsilon \cdot \tau)(\omega) = p_\upsilon(\omega \sigma_{KG}(\tau))$$

$$= p_\upsilon(\omega \tau^{-1})$$

$$= p_{\upsilon \tau}(\omega).$$

From Example 3.2.8, KG^* is a right KG-comodule with structure map $\Psi : KG^* \to KG^* \otimes KG$ defined as $\Psi(p_\upsilon) = p_\upsilon \otimes \upsilon$. Now, KG^* is a right Hopf module over KG, that is, Ψ is a KG-module homomorphism with KG-module structure on $KG^* \otimes KG$ given as

$$(p_\omega \otimes \tau)\upsilon = p_\omega \cdot \upsilon \otimes \tau \upsilon = p_{\omega \upsilon} \otimes \tau \upsilon.$$

We now proceed with the proof of the case $M = H^*$ of the Fundamental Theorem of Hopf Modules. Here is our outline for the proof.

Step 1.   We show that

$$\int_{H^*}^l = W = \{\beta \in H^* : \Psi(\beta) = \beta \otimes 1\}.$$

Step 2.   We give $\int_{H^*}^l$ the trivial right H-module structure and endow $\int_{H^*}^l \otimes H$ with the trivial right Hopf module structure over H.

Step 3.   We show that $H^* \cong \int_{H^*}^l \otimes H$ as right Hopf modules over H.

Proposition 3.2.18. *Step 1.* $\int_{H^*}^l = W$ *where* $W = \{\beta \in H^* : \Psi(\beta) = \beta \otimes 1\}.$

Proof. Let $\{b_i\}_{i=1}^n$ be a basis for H and let $\{\alpha_i\}$ be the basis for H^* dual to $\{b_i\}$. Let $\beta \in \int_{H^*}^l$. Then for all $\alpha \in H^*$,

$$\alpha\beta = \epsilon_{H^*}(\alpha)\beta = \alpha(1)\beta.$$

Thus,

$$\alpha_i\beta = \alpha_i(1)\beta.$$

Thus,

$$\Psi(\beta) = \sum_{i=1}^n \alpha_i\beta \otimes b_i$$

$$= \sum_{i=1}^n \alpha_i(1)\beta \otimes b_i$$

$$= \sum_{i=1}^n \beta \otimes \alpha_i(1)b_i$$

$$= \beta \otimes 1,$$

and so, $\beta \in W$. Conversely, if $\beta \in W$, then for all $\alpha \in H^*$,

$$\alpha\beta = \alpha(1)\beta = \epsilon_{H^*}(\alpha)\beta$$

by (3.9), and so, $\beta \in \int_{H^*}^l$. $\square$

Step 2.   We give $\int_{H^*}^l$ the trivial right H-module structure

$$\beta \cdot h = \epsilon_H(h)\beta,$$

for all $\beta \in \int_{H^*}^l$, $h \in H$, give $\int_{H^*}^l \otimes H$ the structure of a right H-module through

$$(\alpha \otimes k)h = \alpha \otimes kh,$$

for $\alpha \in \int_{H^*}^l$, $k, h \in H$, and endow $\int_{H^*}^l \otimes H$ with the trivial right H-Hopf module structure

$$(I_{\int_{H^*}^l} \otimes \Delta_H) : \int_{H^*}^l \otimes H \to \int_{H^*}^l \otimes H \otimes H.$$

Step 3. Finally, we show that $H^* \cong \int_{H^*}^l \otimes H$ as right H-Hopf modules. We first prove two lemmas.

For $\alpha \in H^*$, let

$$\rho(\alpha) = \sum_{(\alpha)} \alpha_{(1)} \cdot \sigma(a_{(2)}) \in H^*,$$

with $\Psi(\alpha) = \sum_{(\alpha)} \alpha_{(1)} \otimes a_{(2)}$.

Lemma 3.2.19. *For $\alpha \in H^*$, $h \in H$, $\rho(\alpha \cdot h) = \epsilon(h)\rho(\alpha)$.*

Proof. By Proposition 3.2.16

$$\Psi(\alpha \cdot h) = \Psi(\alpha)h = \sum_{(\alpha, h)} (\alpha_{(1)} \cdot h_{(1)}) \otimes a_{(2)} h_{(2)}.$$

Thus

$$\rho(\alpha \cdot h) = \sum_{(\alpha, h)} (\alpha_{(1)} \cdot h_{(1)}) \cdot \sigma(a_{(2)} h_{(2)})$$

$$= \sum_{(\alpha, h)} (\alpha_{(1)} \cdot h_{(1)}) \cdot \sigma(h_{(2)}) \sigma(a_{(2)})$$

$$= \sum_{(\alpha, h)} \alpha_{(1)} \cdot ((h_{(1)}) \sigma(h_{(2)}) \sigma(a_{(2)}))$$

$$= \epsilon(h) \sum_{(\alpha)} \alpha_{(1)} \cdot \sigma(a_{(2)}) \quad \text{by the coinverse property}$$

$$= \epsilon(h)\rho(\alpha).$$

$\square$

Lemma 3.2.20. *For $\alpha \in H^*$, $\Psi(\rho(\alpha)) = \rho(\alpha) \otimes 1$.*

Proof. We have

$$\Psi(\rho(\alpha)) = \Psi\left(\sum_{(\alpha)} \alpha_{(1)} \cdot \sigma(a_{(2)})\right)$$

$$= \sum_{(\alpha)} \Psi(\alpha_{(1)}) \sigma(a_{(2)}) \quad \text{by Proposition 3.2.16}$$

$$= \sum_{(\alpha)} (\alpha_{(1)} \otimes a_{(2)}) \sigma(a_{(3)})$$

$$= \sum_{(\alpha)} (\alpha_{(1)} \cdot \sigma(a_{(4)})) \otimes a_{(2)} \sigma(a_{(3)}) \quad \text{by Proposition 3.1.10(i)}$$

$$= \sum_{(\alpha)} (\alpha_{(1)} \cdot \sigma(a_{(3)})) \otimes \epsilon(a_{(2)})1 \quad \text{by the coinverse property}$$

$$= \left(\sum_{(\alpha)} (\alpha_{(1)} \cdot \sigma(\epsilon(a_{(2)})a_{(3)})) \right) \otimes 1$$

$$= \left(\sum_{(\alpha)} (\alpha_{(1)} \cdot \sigma(a_{(2)})) \right) \otimes 1 \quad \text{by the counit property}$$

$$= \rho(\alpha) \otimes 1.$$

$\square$

By Lemma 3.2.20, $\Psi(\rho(H^*)) = \rho(H^*) \otimes 1$. Thus by Proposition 3.2.18, $\rho(H^*) \subseteq \int_{H^*}^l$, and so, there exists a map $\varrho : H^* \to \int_{H^*}^l \otimes H$ defined as

$$\varrho(\alpha) = (\rho \otimes I_H)\Psi(\alpha) = \sum_{(\alpha)} \rho(\alpha_{(1)}) \otimes a_{(2)}.$$

Proposition 3.2.21 (Larson and Sweedler, [LS69]). *Let H^* be a right H-Hopf module as in Proposition 3.2.16 and let $\int_{H^*}^l \otimes H$ be a trivial right H-Hopf module as in Example 3.2.13. Then $H^* \cong \int_{H^*}^l \otimes H$ as right H-Hopf modules.*

Proof. Define a map $\varphi : \int_{H^*}^l \otimes H \to H^*$ by

$$\varphi(\alpha \otimes h) = \alpha \cdot h,$$

for $\alpha \in \int_{H^*}^l, h \in H$. The map φ is an H-module homomorphism:

$$\varphi((\alpha \otimes h)k) = \varphi(\alpha \otimes hk)$$
$$= \alpha \cdot (hk)$$
$$= (\alpha \cdot h) \cdot k$$
$$= \varphi(\alpha \cdot h)k.$$

We show that φ is an isomorphism of H-modules by showing that $\varrho\varphi = I_{\int_{H^*}^l \otimes H}$ and that $\varphi\varrho = I_{H^*}$. Let $h \in H, \alpha \in \int_{H^*}^l$. We have

$$\varrho\varphi(\alpha \otimes h) = \varrho(\alpha \cdot h)$$
$$= \sum_{(h)} \rho(\alpha \cdot h_{(1)}) \otimes h_{(2)} \quad \text{since } \Psi(\alpha \cdot h) = \alpha \cdot h_{(1)} \otimes h_{(2)}$$

$$= \sum_{(h)} \epsilon(h_{(1)})\rho(\alpha) \otimes h_{(2)} \quad \text{by Lemma 3.2.19}$$

$$= \sum_{(h)} \rho(\alpha) \otimes \epsilon(h_{(1)})h_{(2)}$$

$$= \rho(\alpha) \otimes h \quad \text{by the counit property.}$$

Now by Proposition 3.2.18, $\Psi(\alpha) = \alpha \otimes 1$, and so, $\rho(\alpha) = \alpha$. It follows that

$$\varrho\varphi(\alpha \otimes h) = \alpha \otimes h,$$

thus $\varrho\varphi$ is the identity on $\int_{H^*}^l \otimes H$.

Regarding the map $\varphi\varrho : H^* \to H^*$,

$$\varphi\varrho(\alpha) = \varphi\left(\sum_{(\alpha)} \rho(\alpha_{(1)}) \otimes a_{(2)} \right)$$

$$= \varphi\left(\sum_{(\alpha)} (\alpha_{(1)} \cdot \sigma(a_{(2)})) \otimes a_{(3)} \right)$$

$$= \sum_{(\alpha)} (\alpha_{(1)} \cdot \sigma(a_{(2)})) \cdot a_{(3)})$$

$$= \sum_{(\alpha)} \alpha_{(1)} \cdot \sigma(a_{(2)})a_{(3)}$$

$$= \sum_{(\alpha)} \alpha_{(1)} \cdot \epsilon(a_{(2)})1 \quad \text{by the coinverse property}$$

$$= \sum_{(\alpha)} \epsilon(a_{(2)})\alpha_{(1)}$$

$$= \alpha$$

since H^* is a right H-comodule. Thus $\varphi\varrho$ is the identity on H^*. And so φ is an isomorphism of H-modules.

Finally, we show that φ is a homomorphism of right H-comodules, that is, we show that φ preserves right H-comodule structure. For $\alpha \otimes h \in \int_{H^*}^l \otimes H$, we have

$$\Psi(\varphi(\alpha \otimes h)) = \Psi(\alpha \cdot h)$$

$$= \Psi(\alpha)h$$

$$= (\alpha \otimes 1)h$$

$$= \sum_{(h)} \alpha \cdot h_{(1)} \otimes h_{(2)}$$

$$= \sum_{(h)} \varphi(\alpha \otimes h_{(1)}) \otimes h_{(2)}$$

$$= (\varphi \otimes I_H)\left(\sum_{(h)} \alpha \otimes h_{(1)} \otimes h_{(2)} \right)$$

$$= (\varphi \otimes I_H)(I_{\int_{H^*}^l} \otimes \Delta_H)(\alpha \otimes h),$$

and so, φ is an isomorphism of right H-Hopf modules.   □

Example 3.2.22. Let $H = KG$ and $H^* = KG^*$. By Proposition 3.2.5, $\int_{KG^*}^l = Kp_1$. As in Proposition 3.2.21, there is an isomorphism of right KG-Hopf modules φ : $Kp_1 \otimes KG \to KG^*$ defined by

$$\varphi(p_1 \otimes \upsilon) = p_1 \cdot \upsilon = p_\upsilon.$$

Its inverse $\varrho : KG^* \to Kp_1 \otimes KG$ is defined by

$$\varrho(p_\upsilon) = (\rho \otimes I_{KG})\Psi(p_\upsilon)$$

$$= (\rho \otimes I_{KG})(p_\upsilon \otimes \upsilon)$$

$$= \rho(p_\upsilon) \otimes \upsilon$$

$$= (p_\upsilon \cdot \sigma(\upsilon)) \otimes \upsilon$$

$$= (p_\upsilon \cdot \upsilon^{-1}) \otimes \upsilon$$

$$= p_1 \otimes \upsilon.$$

Corollary 3.2.23. *Let H be an n-dimensional K-Hopf algebra with space of left integrals $\int_H^l$. Then $\dim(\int_H^l) = 1$. Consequently, there exists an integral $\Lambda \in \int_H^l$ for which $K\Lambda = \int_H^l$.*

Proof. By Proposition 3.2.21, with H in place of H^*, there exists an isomorphism of H^*-modules, and thus of K-vector spaces

$$\int_H^l \otimes H^* \cong H.$$

Consequently, $\dim(\int_H^l)\dim(H^*) = \dim(H)$. But $n = \dim(H^*) = \dim(H)$, and so $\dim(\int_H^l) = 1$. Thus there exists an integral $\Lambda \in \int_H^l$ for which $K\Lambda = \int_H^l$.   □

An integral $\Lambda \in \int_H^l$ for which $K\Lambda = \int_H^l$ is a **generating integral** for $\int_H^l$. As we have seen, for G a finite group, $\Lambda = \sum_{\tau \in G} \tau$ is a generating integral for KG and $\Lambda = p_1$ is a generating integral for KG^*.

3.3 Hopf Algebras over Rings

In this section we define Hopf algebras over a ring R. Many notions of Hopf algebras over fields generalize to rings, including integrals, comodules, and Hopf modules. If H is a Hopf algebra of finite rank over R, then there is an isomorphism $H^* \cong \int_{H^*}^l \otimes_R H$ as right H-Hopf modules; if R is also a PID, then one can show that H^* admits a generating integral.

*   *   *

Let R be a commutative ring with unity.

Definition 3.3.1. A **Hopf algebra H over R** is an R-algebra H with structure map $\lambda_H : R \to H$, together with additional R-linear maps

$$\Delta_H : H \to H \otimes_R H, \quad \text{(comultiplication)},$$

$$\epsilon_H : H \to R, \quad \text{(counit)},$$

$$\sigma_H : H \to H, \quad \text{(coinverse)}.$$

The maps Δ_H, ϵ_H and σ_H satisfy the identical comultiplication, counit, and coinverse properties, respectively, and the maps Δ_H and ϵ_H are R-algebra homomorphisms.

The obvious example of an R-Hopf algebra is the group ring RG where G is any finite group. As we shall see, in the case that R is an integral domain with field of fractions K, an R-Hopf order in KG is also an R-Hopf algebra.

Many of the results for Hopf algebras over K carry over to rings. For example, if H is a free R-module of finite rank n, then so is its linear dual $H^* = \text{Hom}_R(H, R)$. The ideal of left integrals $\int_{H^*}^l$ is defined as before:

$$\int_{H^*}^l = \{y \in H^* : xy = e_{H^*}(x)y, \forall x \in H^*\}.$$

In fact, Proposition 3.2.21 holds: $\int_{H^*}^l \otimes_R H \cong H^*$ as right H-Hopf modules. If R is a PID, then the R-submodule $\int_{H^*}^l \subseteq H^*$ is free of rank $l \leq n$. We conclude that $\int_{H^*}^l$ is a free rank one R-module, that is, there is a generating integral for H^*.

3.4 Hopf Orders

In this section we introduce Hopf orders in the group ring Hopf algebra KG. We show that Hopf orders are Hopf algebras over rings with structure maps induced from those of the K-Hopf algebra KG. When R is the ring of integers of a finite extension $K/\mathbb{Q}$, we construct a collection of one parameter Hopf orders $H(i)$ in KC_p

together with their linear duals, $H(i)^*$. In §4.5, we show how Hopf orders can be used to generalize the concept of a Galois group.

*   *   *

Let R be an integral domain, let K be its field of fractions, and assume that R is integrally closed in K. Let G be a finite group of order n. The group ring KG is a K-Hopf algebra with comultiplication $\Delta_{KG} : KG \to KG \otimes_K KG$ defined by $g \mapsto g \otimes g$, counit $\epsilon_{KG} : KG \to K$, defined by $g \mapsto 1$, and coinverse $\sigma_{KG} : KG \to KG$ given by $g \mapsto g^{-1}$, for $g \in G$.

Definition 3.4.1. An R-**order in** KG is an R-submodule A of KG that satisfies the conditions

(i) A is finitely generated and projective as an R-module,
(ii) $A \subseteq KG$ is closed under the multiplication of KG and $1_{KG} \in A$,
(iii) A contains a K-basis for KG (equivalently, $KA = KG$).

Proposition 3.4.2. *Let A be an R-order in KG. Then every element of A is a zero of a monic polynomial with coefficients in R.*

Proof. Let $\alpha \in A$. Since A is an R-order, $\alpha A \subseteq A$. This implies that α satisfies a monic polynomial with coefficients in R (use the "Integrality Theorem" [La84, IX, §1]). □

Proposition 3.4.3. *Let R be a local integrally closed integral domain with field of fractions K. Let A be an R-order in KG. Then A is free over R of rank $|G|$.*

Proof. By definition, A is finitely generated and projective as an R-module, and so A is free over R of rank, say, m. Moreover, $m = |G|$ since $KA = KG$. □

Definition 3.4.4. An R-order H in KG for which $\Delta_{KG}(H) \subseteq H \otimes H$ is an R-**Hopf order in** KG.

For example, the group ring RG is an R-Hopf order in KG. When $G = 1$, then the Hopf order $R1 = R$ in $K1 = K$ is the **trivial Hopf order**.

Proposition 3.4.5. *Let H be an R-Hopf order in KG where G is a finite group of order $n \geq 1$. Then the coinverse σ_{KG} and counit ϵ_{KG} satisfy:*

(i) $\sigma_{KG}(H) = H$,
(ii) $\epsilon_{KG}(H) = R$.

Proof. For (i): Let $m = m_{KG}$, $I = I_{KG}$, $\Delta = \Delta_{KG}$. For $n = 1$ or $n = 2$, σ_{KG} is the identity map on KG and so (i) holds. For $n \geq 3$, let

$$m^{(n-2)} = m(I \otimes m)(I \otimes I \otimes m) \cdots \underbrace{(I \otimes I \cdots \otimes I}_{n-3} \otimes m),$$

$$\Delta^{(n-2)} = \underbrace{(I \otimes I \cdots \otimes I}_{n-3} \otimes \Delta) \cdots (I \otimes I \otimes \Delta)(I \otimes \Delta)\Delta.$$

Then for all $g \in G$,

$$m^{(n-2)} \Delta^{(n-2)}(g) = g^{n-1} = g^{-1},$$

and so,

$$\sigma_{KG} = m^{(n-2)} \Delta^{(n-2)}.$$

Now,

$$\sigma_{KG}(H) = m^{(n-2)}(\Delta^{(n-2)}(H)) \subseteq H,$$

since $m(H \otimes H) \subseteq H$ and $\Delta(H) \subseteq H \otimes H$, thus $\sigma_{KG}(H) \subseteq H$. Since $\sigma_{KG}^2 = I_{KG}$, $H \subseteq \sigma_{KG}(H)$. This proves (i).

For (ii): Recall that the K-algebra structure map $\lambda_{KG} : K \to KG$ is defined by $\lambda_{KG}(r) = r$, for $r \in K$. Thus $1_K = 1_{KG}$ in KG. The R-module structure of H is defined through the restriction of λ_{KG} to R, and so $1_R \in H$. Thus $R \subseteq H$, and consequently, $R = \epsilon_{KG}(R) \subseteq \epsilon_{KG}(H)$.

We next show that $\epsilon_{KG}(H) \subseteq R$. For $h \in H$,

$$m_{KG}(I_{KG} \otimes \sigma_{KG})\Delta_{KG}(h) = \epsilon_H(h)1_{KG} = \epsilon_H(h).$$

Consequently, $\epsilon_{KG}(H) \subseteq H \cap K$. Since H is finitely generated as an R-module, $\epsilon_{KG}(H)$ is finitely generated as an R-submodule of K. Moreover, for $a, b \in H$,

$$\epsilon_{KG}(a)\epsilon_{KG}(b) = \epsilon_{KG}(ab) \in \epsilon_{KG}(H),$$

since H is closed under the multiplication of KG. Thus $\epsilon_{KG}(H)$ is closed under the multiplication in K. Let $s \in \epsilon_{KG}(H)$. By [La84, IX, §1] s is integral over R, and since R is integrally closed, $s \in R$. Thus $\epsilon_{KG}(H) \subseteq R$, and so $\epsilon_{KG}(H) = R$. ☐

Proposition 3.4.6. *Let H be an R-Hopf order in KG, G a finite group. Then H is an R-Hopf algebra.*

Proof. H is a ring with multiplication m_{KG} since H is closed under the multiplication of KG. Moreover, as an R-submodule of KG, the R-module structure of H is given by the restriction of λ_{KG} to R. Thus, H is an R-algebra. Since H is closed under the comultiplication of KG, the required comultiplication Δ_H can be taken to be Δ_{KG} restricted to H. By Proposition 3.4.5(ii), the counit map ϵ_H can be taken to be ϵ_{KG} restricted to H, and by Proposition 3.4.5(i), the coinverse map σ_H is σ_{KG} restricted to H. Note that Δ_H, ϵ_H, and σ_H satisfy the comultiplication, counit, and coinverse properties, respectively, since KG is a K-Hopf algebra. ☐

We describe a collection of R-Hopf orders in KG. Let p be a prime number, let $G = C_p$ denote the cyclic group of order p generated by g. For an integer $n \geq 1$, let

$K = \mathbb{Q}(\zeta_{p^n})$ where ζ_{p^n} is a primitive p^nth root of unity. Then the ring of integers R of K is $\mathbb{Z}[\zeta_{p^n}]$, and the ideal (p) factors uniquely as

$$(p) = (1 - \zeta_{p^n})^{p^{n-1}(p-1)}.$$

Put $\lambda = 1 - \zeta_{p^n}$. For an integer $0 \le i \le p^{n-1}$, let $H(i)$ denote the free R-module on the basis

$$\left\{ 1, \left(\frac{g-1}{\lambda^i}\right), \left(\frac{g-1}{\lambda^i}\right)^2, \ldots, \left(\frac{g-1}{\lambda^i}\right)^{p-1} \right\}. \tag{3.13}$$

Proposition 3.4.7. *For each integer $0 \le i \le p^{n-1}$, $H(i)$ is an R-Hopf order in KC_p.*

Proof. We first show that $H(i)$ is an R-order in KC_p. Since $H(i)$ is free over R of rank p, it is certainly finitely generated and projective as an R-module. Also, $1 = 1_{KG} \in H(i)$ by construction. Put $h = \frac{g-1}{\lambda^i}$. Then $g = \lambda^i h + 1$ so that

$$1 = g^p$$
$$= (\lambda^i h + 1)^p$$
$$= \lambda^{pi} h^p + \binom{p}{1} \lambda^{(p-1)i} h^{p-1} + \cdots + \binom{p}{p-1} \lambda^i h + 1.$$

Hence

$$\lambda^{pi} h^p + \binom{p}{1} \lambda^{(p-1)i} h^{p-1} + \cdots + \binom{p}{p-1} \lambda^i h = 0,$$

so that

$$h^p = -\binom{p}{1} \lambda^{-i} h^{p-1} - \cdots - \binom{p}{p-1} \lambda^{-(p-1)i} h.$$

Now $-\binom{p}{m} \lambda^{-mi} \in R$ for $1 \le m \le p-1$ and $0 \le i \le p^{n-1}$ since $p \mid \binom{p}{m}$ for $1 \le m \le p-1$ and $\lambda^{mi} \mid p$ for all $1 \le m \le p-1$, $0 \le i \le p^{n-1}$. It follows that $H(i)$ is closed under the multiplication of KC_p. Also, $KH(i) = KC_p$, and so, $H(i)$ is an R-order in KC_p.

Now

$$\Delta_{KC_p}\left(\frac{g-1}{\lambda_i}\right) = \frac{1}{\lambda^i}(g \otimes g - 1 \otimes 1)$$

$$= \frac{1}{\lambda^i}((g - 1 + 1) \otimes (g - 1 + 1) - 1 \otimes 1)$$

$$= \frac{1}{\lambda^i}((g - 1) \otimes (g - 1) + (g - 1) \otimes 1$$

$$+ 1 \otimes (g - 1) + 1 \otimes 1 - 1 \otimes 1)$$

$$= \frac{1}{\lambda^i}(1 \otimes (g-1) + (g-1) \otimes 1 + (g-1) \otimes (g-1))$$

$$= 1 \otimes \left(\frac{g-1}{\lambda^i}\right) + 1 \otimes \left(\frac{g-1}{\lambda^i}\right) + \lambda^i \left(\frac{g-1}{\lambda^i}\right) \otimes \left(\frac{g-1}{\lambda^i}\right).$$

Thus $\Delta_{KC_p}\left(\frac{g-1}{\lambda^i}\right) \in H(i) \otimes H(i)$. Since Δ_{KC_p} is an algebra homomorphism and $H(i) \otimes H(i)$ is closed under the multiplication in $KC_p \otimes KC_p$, we conclude that $\Delta_{KC_p}(H(i)) \subseteq H(i) \otimes H(i)$. Thus $H(i)$ is an R-Hopf order in KC_p. $\square$

Note that $H(i)$ is generated as an R-algebra by $\left\{\frac{g-1}{\lambda^i}\right\}$, thus

$$H(i) = R\left[\frac{g-1}{\lambda^i}\right].$$

Observe that $H(0) = R[g-1] = R[g] = RC_p$.

Put $\zeta_p = \zeta_{p^n}^{p^{n-1}}$. Let $\hat{C}_p$ denote the character group of C_p generated by γ, with $\gamma^l(g^m) = \zeta_p^{lm}$, for $0 \le l, m \le p-1$. Let tr $: KC_p \to K$ denote the trace map defined as

$$\mathrm{tr}(x) = \sum_{l=0}^{p-1} \gamma^l(x).$$

Note that $\mathrm{tr}\left(\sum_{m=0}^{p-1} a_m g^m\right) = p a_0$ for $a_m \in K$. As one can check, the map $B : KC_p \times KC_p \to K$ defined as $B(x, y) = \mathrm{tr}(xy)$ is a symmetric non-degenerate bilinear form on KC_p, cf. [CF67, Chapter 1, §3]. With respect to B, we compute the discriminant of RC_p, denoted as $\mathrm{disc}(RC_p)$.

Proposition 3.4.8. *With respect to B, $\mathrm{disc}(RC_p) = (p)^p = (\lambda^{p^n(p-1)})$.*

Proof. Using the basis $\{1, g, g^2, \ldots, g^{p-1}\}$ for RC_p, one computes $B(g^m, g^n) = \mathrm{tr}(g^{m+n}) = p\delta_{0,m+n}$, where $m+n$ is taken modulo p. The result follows. $\square$

We next compute $\mathrm{disc}(H(i))$ for $0 \le i \le p^{n-1}$.

Proposition 3.4.9. *For $0 \le i \le p^{n-1}$, $\mathrm{disc}(H(i)) = (\lambda^{p(p-1)(p^{n-1}-i)})$.*

Proof. Clearly, $H(0) \subseteq H(i)$ for all i. We first compute the module index $[H(i) : H(0)]$. We find the matrix T in $\mathrm{Mat}_p(K)$ that multiplies the R-basis

$$\left\{1, \left(\frac{g-1}{\lambda^i}\right), \left(\frac{g-1}{\lambda^i}\right)^2, \ldots, \left(\frac{g-1}{\lambda^i}\right)^{p-1}\right\},$$

for $H(i)$ to yield the R-basis

$$\left\{1, (g-1), (g-1)^2, \ldots, (g-1)^{p-1}\right\}$$

for $H(0)$. One has

$$T = \begin{pmatrix} 1 & 0 & 0 & \cdots & 0 \\ 0 & \lambda^i & 0 & \cdots & 0 \\ 0 & 0 & \lambda^{2i} & \cdots & 0 \\ \vdots & & & \ddots & \vdots \\ 0 & 0 & 0 & \cdots & \lambda^{(p-1)i} \end{pmatrix}$$

Thus

$$\begin{aligned} [H(i) : H(0)] &= R \det(T) \\ &= R\lambda^i \lambda^{2i} \cdots \lambda^{(p-1)i} \\ &= R\lambda^{i+2i+\cdots+(p-1)i} \\ &= (\lambda^{p(p-1)i/2}). \end{aligned}$$

Now,

$$\begin{aligned} \operatorname{disc}(H(0)) &= [H(i) : H(0)]^2 \operatorname{disc}(H(i)) \\ &= (\lambda^{p(p-1)i})\operatorname{disc}(H(i)), \end{aligned}$$

and so,

$$\begin{aligned} \operatorname{disc}(H(i)) &= (\lambda^{-p(p-1)i})\operatorname{disc}(H(0)) \\ &= (\lambda^{-p(p-1)i})(\lambda^{p^n(p-1)}) \quad \text{by Proposition 3.4.8} \\ &= (\lambda^{p(p-1)(p^{n-1}-i)}). \end{aligned}$$

$\square$

The dual module of $H(i)$ is defined as

$$H(i)^D = \{x \in KC_p : B(x, H(i)) \subseteq R\}.$$

We want to compute $\operatorname{disc}(H(i)^D)$, for $0 \le i \le p^{n-1}$, and ultimately, show that $H(i)^D$ is an R-Hopf order in KC_p. We consider the $i = 0$ case first. Since

$$B\left(\frac{g^{-m}}{p}, g^n\right) = \operatorname{tr}(g^{-m+n}/p) = \delta_{m,n},$$

$$\left\{ \frac{1}{p}, \frac{g^{-1}}{p}, \frac{g^{-2}}{p}, \dots, \frac{g^{-(p-1)}}{p} \right\},$$

is the R-basis for $H(0)^D$ dual to the R-basis $\{1, g, g^2, \ldots, g^{p-1}\}$ of RC_p. We identify $H(0)^D$ with the collection of R-linear maps $\text{Hom}_R(H(0), R)$: for $x \in H(0)^D$, $a \in H(0)$, $x(a) = B(x, a)$.

Now, the R-submodule $H(0)^D$ of KC_p is an R-algebra with multiplication (again, convolution!) defined as

$$(x * y)(a) = (B(x, -) \otimes B(y, -))\Delta_{KC_p}(a)$$

$$= \sum_{(a)} B(x, a_{(1)})B(y, a_{(2)})$$

$$= \sum_{(a)} \text{tr}(xa_{(1)})\text{tr}(ya_{(2)}),$$

for $a \in H(0)$, $x, y \in H(0)^D$. For $g^l \in G$, $0 \le m, n \le p - 1$,

$$\left(\frac{g^{-m}}{p} * \frac{g^{-n}}{p}\right)(g^l) = (B(g^{-m}/p, -) \otimes B(g^{-n}/p, -))\Delta_{KC_p}(g^l)$$

$$= B(g^{-m}/p, g^l)B(g^{-n}/p, g^l)$$

$$= \text{tr}(g^{-m+l}/p)\text{tr}(g^{-n+l}/p)$$

$$= \delta_{m,l}\delta_{n,l}$$

$$= \delta_{m,n}\delta_{n,l}$$

$$= \left(\delta_{m,n}\frac{g^{-n}}{p}\right)(g^l).$$

Thus $\left\{\frac{g^{-m}}{p}\right\}$, $0 \le m \le p - 1$, is the collection of minimal idempotents in the R-algebra $H(0)^D$.

The minimal idempotents in the K-algebra KC_p is the set $\{e_0, e_1, \ldots, e_{p-1}\}$ with

$$e_m = \frac{1}{p}\sum_{n=0}^{p-1}\zeta_p^{-nm}g^n$$

for $0 \le m \le p - 1$. Consequently, there is an isomorphism of R-algebras

$$\phi : H(0)^D \to \bigoplus_{m=0}^{p-1} Re_m,$$

defined by $\phi\left(\frac{g^{-m}}{p}\right) = e_m$. Note that $e_m \in H(0)^D$ for $0 \le m \le p - 1$ and $e_m * e_n = e_{m+n}$, $m + n$ taken modulo p. Thus $\phi(e_m) = g^m$.

We identify $H(0)^D$ with the R-order $\bigoplus_{m=0}^{p-1} Re_m$ in KC_p through ϕ, hence, an R-basis for $H(0)^D$ is $\{e_0, e_1, \ldots, e_{p-1}\}$. One has

$$B(e_m, e_n) = \mathrm{tr}(e_m e_n) = \mathrm{tr}(\delta_{m,n} e_n) = \delta_{m,n},$$

and so, $\mathrm{disc}(H(0)^D) = R$. We can now compute $\mathrm{disc}(H(i)^D)$ for $0 \leq i \leq p^{n-1}$.

Proposition 3.4.10. *Let* $0 \leq i \leq p^{n-1}$ *and let* $H(i)$ *be an* R-*Hopf order in* KC_p. *Then* $\mathrm{disc}(H(i)^D) = (\lambda^{p(p-1)i})$.

Proof. Since $H(0) \subseteq H(i)$, $H(i)^D \subseteq H(0)^D = \bigoplus_{m=0}^{p-1} Re_m$. Now,

$$
\begin{aligned}
\mathrm{disc}(H(i)^D) &= [H(0)^D : H(i)^D]^2 \mathrm{disc}(H(0)^D) \\
&= [H(0)^D : H(i)^D]^2 \\
&= [H(i) : H(0)]^2 \\
&= (\lambda^{p(p-1)i}).
\end{aligned}
$$

$\square$

Finally, we show that $H(i)^D$ is an R-Hopf order in KC_p of the form $H(j)$ for some j, $0 \leq j \leq p^{n-1}$.

Proposition 3.4.11. *Let* $0 \leq i \leq p^{n-1}$, *let* $i' = p^{n-1} - i$, *and let* $H(i)$ *be an* R-*Hopf order in* KC_p. *Then* $H(i)^D$ *is an* R-*Hopf order in* KC_p *with* $H(i)^D = H(i')$, *specifically,* $H(0)^D = H(p^{n-1})$.

Proof. Since $H(0) \subseteq H(i)$, $H(i)^D \subseteq H(0)^D$. Endow $H(0)^D$ with convolution $*$ as multiplication. For $a \in H(0)^D$, let

$$a^{*^l} = \underbrace{a * a * \cdots * a}_{l}.$$

Let $e_0, e_1, e_2, \ldots, e_{p-1}$ be the elements in $H(0)^D$ defined as $e_m = \frac{1}{p} \sum_{n=0}^{p-1} \zeta_p^{-nm} g^n$. Let A be the free R-module on the basis

$$\left\{ e_0, \left(\frac{e_1 - e_0}{\lambda^{i'}}\right), \left(\frac{e_1 - e_0}{\lambda^{i'}}\right)^{*2}, \ldots, \left(\frac{e_1 - e_0}{\lambda^{i'}}\right)^{*p-1} \right\}.$$

Now,

$$B\left(e_0, \left(\frac{g-1}{\lambda^i}\right)^m \right) = \delta_{0,m},$$

for $0 \leq m \leq p - 1$,

$$B\left(\left(\frac{e_1 - e_0}{\lambda^{i'}}\right)^{*^l}, 1 \right) = \delta_{l,0},$$

for $1 \le l \le p-1$, and

$$B\left(\left(\frac{e_1 - e_0}{\lambda^{i'}}\right)^{*l}, \left(\frac{g-1}{\lambda^i}\right)^m\right) = \left(\frac{\zeta_p - 1}{\lambda^{p^{n-1}}}\right)^{lm} \in R,$$

for $1 \le l, m \le p-1$. Thus, $A \subseteq H(i)^D$.

Since $\phi(e_l) = g^l$, we have

$$\phi\left(\left(\frac{e_1 - e_0}{\lambda^{i'}}\right)^{*l}\right) = \left(\frac{g-1}{\lambda^{i'}}\right)^l,$$

for $1 \le l \le p-1$, thus $\phi(A) = H(i')$, that is, $A \cong H(i')$, as R-algebras. We make the identification $A = H(i')$, thus $H(i') \subseteq H(i)^D$. Now,

$$\text{disc}(H(i')) = \left(\lambda^{p(p-1)(p^{n-1}-i')}\right), \quad \text{by Proposition 3.4.9}$$

$$= \left(\lambda^{p(p-1)(p^{n-1}-(p^{n-1}-i))}\right)$$

$$= \left(\lambda^{p(p-1)i}\right)$$

$$= \text{disc}(H(i)^D) \quad \text{by Proposition 3.4.10.}$$

It follows that $H(i') = H(i)^D$. In the case $i = 0$, $H(0)^D = H(p^{n-1})$. $\qquad\square$

We next consider Hopf orders over a local ring. We fix a prime number p and an integer $n \ge 1$ and assume that K is a finite extension of $\mathbb{Q}$ with ring of integers R containing a primitive p^nth root of unit ζ_{p^n}. Let

$$(p) = P_1^{e_1} P_2^{e_2} \cdots P_m^{e_m}$$

be the unique factorization of (p) into prime ideals of R. Take $P = P_i$ and $e = e_i$ for some i and let $|\ |_P$ denote the corresponding absolute value on K. There exists a field extension L of K with the following properties:

(i) The absolute value $|\ |_P$ extends uniquely to an absolute value on L, also denoted as $|\ |_P$,

(ii) with respect to $|\ |_P$, K is dense in L, that is, the closure $\overline{K} = L$,

(iii) L is complete with respect to $|\ |_P$, that is, every $|\ |_P$-Cauchy sequence in L converges to an element in L.

The field extension L is the **completion of K with respect to $|\ |_P$**, or: the completion of K at the prime ideal P, and is denoted as K_P. Note that K_P is a finite extension of $\mathbb{Q}_p$ of local degree $[K_P : \mathbb{Q}_p]$.

Let π be the uniformizing parameter for K_P, and let R_P be the valuation ring of K_P. Then

$$(p) = (\pi)^e.$$

We have $p = u\pi^e$ for some unit $u \in R_P$, and so $\mathrm{ord}_\pi(p) = e$. Since $p = v(1-\zeta_p)^{p-1}$ for some unit v,

$$e = \mathrm{ord}_\pi(p) = (p-1)\,\mathrm{ord}_\pi(1 - \zeta_p).$$

Thus $e/(p-1)$ is an integer which we denote as e'; we have $\mathrm{ord}_\pi(1 - \zeta_p) = e'$. For instance, in the extension $\mathbb{Q}(\zeta_{p^n})/\mathbb{Q}$ of Proposition 3.4.7, $e' = p^{n-1}$.

Let C_{p^n} denote the cyclic group of order p^n generated by g. There is a considerable body of research on the structure of R_P-Hopf orders in $K_P C_{p^n}$, see [TO70, Gr92, By93b, Un94, Un96, CU03, CU04, UC05], and [Un08b]. In fact, Hopf orders in $K_P C_{p^n}$ have been completely classified in the cases $n = 1, 2$, see [TO70, Gr92, By93b] and [Un94]. The classification in the $n = 1$ case is due to Tate and Oort [TO70]. Essentially, an R_P-Hopf order in $K_P C_p$ looks like the local version of a Hopf order constructed in Proposition 3.4.7.

Proposition 3.4.12 (Tate and Oort). *Let H be an R_P-Hopf order in $K_P C_p$. Then*

$$H = R_P\left[\frac{g-1}{\pi^i}\right]$$

for some integer i, $0 \le i \le e'$.

Proof. The proof is beyond the scope of this book. The interested reader is referred to [Ch00, Chapter 4]. □

3.5   Chapter Exercises

Exercises for §3.1

1. Let $K[x]$ be the polynomial bialgebra with x grouplike. Show that there is no K-linear map $K[x] \to K[x]$ that satisfies the coinverse property.
2. Verify that the maps Δ_H, ϵ_H, and σ_H of Example 3.1.5 satisfy the comultiplication, counit, and coinverse properties, respectively.
3. Let H be a cocommutative K-Hopf algebra and let A be a commutative K-algebra. Prove that $f * g = g * f$, for all $f, g \in \mathrm{Hom}_K(H, A)$.
4. Let H be Sweedler's Hopf algebra of Example 3.1.5. Show that the coinverse map σ_H has order 4.
5. Prove Proposition 3.1.7.
6. Let H be a commutative K-Hopf algebra with coinverse σ_H. Prove that $\sigma_H^2 = I_H$.
7. Give an example of a K-Hopf algebra that does not have a bijective coinverse.
8. Let B be a bialgebra over a field K. Let $\sigma_B : B \to B$ be the map defined as $\sigma_B(b) = 0$ for all $b \in B$. Determine whether B together with σ_B is a K-Hopf algebra.

9. Let $S_3 = \langle \sigma, \tau \rangle$, $\sigma^3 = \tau^2 = 1$, $\tau\sigma = \sigma^2\tau$, be the symmetric group on three letters, and let KS_3 be the Hopf algebra of Example 3.1.2. Let p_σ and p_τ be dual basis elements in KS_3^*. Compute $\Delta_{KS_3^*}(p_\sigma)$ and $\Delta_{KS_3^*}(p_\tau)$.

10. Let H be a commutative, cocommutative K-Hopf algebra. Let $m_H\Delta_H : H \to H$ be the map defined as $h \mapsto \sum_{(h)} h_{(1)}h_{(2)}$.

    (a) Show that $m_H\Delta_H : H \to H$ is a homomorphism of K-Hopf algebras.
    (b) In the case that $H = KG$ for G finite, find conditions under which $m_H\Delta_H$ is an isomorphism.

11. Let H be a K-Hopf algebra and view K as the trivial Hopf algebra.

    (a) Prove that $\epsilon_H : H \to K$ is a homomorphism of K-coalgebras.
    (b) Is ϵ_H a homomorphism of bialgebras?
    (c) Is ϵ_H a homomorphism of Hopf algebras?

12. Let H be a K-Hopf algebra. The kernel of ϵ_H is an ideal of H called the **augmentation ideal** of H. Prove that $\ker(\epsilon_H)$ is a Hopf ideal.

13. Let $\phi : H \to H'$ be a homomorphism of Hopf algebras. Suppose that $h \in H$ is grouplike. Show that $\phi(h)$ is grouplike. Suppose that $h \in H$ is primitive. Show that $\phi(h)$ is primitive.

14. Finish the proof of Proposition 3.1.14.

Exercises for §3.2

15. Prove that the Hopf algebra $H = K[x, y]/(xy - 1)$ of Example 3.1.4 is unimodular and compute the ideal of integrals of H.

16. Compute the ideal of left integrals for M. Sweedler's Hopf algebra in Example 3.1.5. Is this Hopf algebra unimodular?

17. Referring to Example 3.2.11, prove that the comultiplication map $\Delta_H : H \to H \otimes H$ is a homomorphism of right H-modules.

18. Referring to Example 3.2.12, prove that $W \otimes H$ is a right H-comodule with structure map $I_W \otimes \Delta_H : W \otimes H \to W \otimes H \otimes H$.

19. Let H be Sweedler's K-Hopf algebra of Example 3.1.5.

    (a) Compute the ideal of left integrals $\int_{H^*}^l$.
    (b) Give explicit definitions for the isomorphisms $\varrho : H^* \to \int_{H^*}^l \otimes H$ and $\varphi : \int_{H^*}^l \otimes H \to H^*$ of Proposition 3.2.21.

20. Let H be a finite dimensional Hopf algebra over a field K. Prove that $H^* \cong \int_{H^*}^l \otimes H$ as vector spaces over K.

Exercises for §3.3

21. Let R be a commutative ring with unity and let H be an R-Hopf algebra.

    (a) Prove that $\epsilon_H(H) = R$.
    (b) Use (a) to prove that $H \cong \ker(\epsilon_H) \oplus R$, as R-modules.

Exercises for §3.4

22. Let $K = \mathbb{Q}(\zeta_{27})$ with ring of integers $R = \mathbb{Z}[\zeta_{27}]$ and let C_3 denote the cyclic group of order 3. Construct the collection of Hopf orders $H(i)$ in KC_3.

23. Let p be a prime number, let C_p denote the cyclic group of order p, and let $K = \mathbb{Q}(\zeta_p)$. Prove that up to isomorphism, the only Hopf orders in KC_p are $\mathbb{Z}[\zeta_p]C_3$ and $\mathbb{Z}[\zeta_p]C_p^D$.

24. Let $K = \mathbb{Q}(\zeta_{p^n})$, $n \geq 1$, with ring of integers $R = \mathbb{Z}[\zeta_{p^n}]$. Then $(p) = (\lambda)^e$, with $e = p^{n-1}(p-1)$, $\lambda = 1 - \zeta_{p^n}$. Let $e' = e/(p-1) = p^{n-1}$, and let i, j be integers $0 \leq i, j \leq e'$ with $pj \leq i$. Let C_{p^2} denote the cyclic group of order p^2 generated by g. Prove that

$$
H(i,j) = R\left[\frac{g^p - 1}{\lambda^i}, \frac{g - 1}{\lambda^j}\right]
$$

is an R-Hopf order in KC_{p^2}.

Questions for Further Study

1. Let K be a finite extension of $\mathbb{Q}$ with ring of integers R. Let KT be the Myhill–Nerode bialgebra of Example 2.2.12.

   (a) Show that RT is an R-order in KT in the sense that the conditions of Definition 3.4.1 are satisfied. Is it true that $\Delta_{KT}(RT) \subseteq RT \otimes RT$?

   (b) Find an example of an R-order A in KT other than RT. Does it hold that $\Delta_{KT}(A) \subseteq A \otimes A$?

2. Let K be a finite extension of $\mathbb{Q}$ with ring of integers R. Let KT be the Myhill–Nerode bialgebra of Example 2.2.13.

   (a) Show that RT is an R-order in KT in the sense that the conditions of Definition 3.4.1 are satisfied. Is it true that $\Delta_{KT}(RT) \subseteq RT \otimes RT$?

   (b) Find an example of an R-order A in KT other than RT. Does it hold that $\Delta_{KT}(A) \subseteq A \otimes A$?

Chapter 4
Applications of Hopf Algebras

In this chapter we present three diverse applications of Hopf algebras. Our first application involves almost cocommutative bialgebras and quasitriangular bialgebras. We show that a quastitriangular bialgebra determines a solution to the Quantum Yang–Baxter Equation, and we give details on how to compute quasitriangular structures for certain two-dimensional bialgebras and Hopf algebras. We show that almost cocommutative Hopf algebras generalize Hopf algebras in which the coinverse has order 2. We then define the braid group on three strands (or more simply, the braid group) and show that a quasitriangular structure determines a representation of the braid group.

For our second application we define affine varieties over a field K and discuss the coordinate ring $K[\Lambda]$ of an affine variety Λ. We show that an affine variety Λ can be identified with the collection of K-algebra maps $K[\Lambda] \to K$, and this allows us to think of the geometric object Λ as an algebraic object through the algebraic structure of its coordinate ring $K[\Lambda]$. We show that if $K[\Lambda]$ is a bialgebra, then Λ is a monoid, and if $K[\Lambda]$ is a Hopf algebra, then Λ is a group.

For our third application, we use Hopf algebras to generalize the concept of a Galois extension. We show that the notion of a Galois extension L of K with group G is equivalent to L being a Galois KG-extension of K. In this latter form (L is a Galois KG-extension of K) the notion of a classical Galois extension L/K can be extended to rings of integers. If S is the ring of integers of L and R is the ring of integers of K, we consider when S is a Galois RG-extension of R, and provide an example of when this occurs using the Hilbert Class Field.

The notion of a Galois KG-(or RG-) extension can be generalized to Galois H-extensions of rings S/R where H is a Hopf algebra. Moreover, the action of the Hopf algebra H on the ring S need not be induced from the classical Galois action. We give a general result of S. Chase and M. Sweedler which yields a Hopf Galois structure on a ring S in which the action of H on S is not the classical Galois action.

© Springer International Publishing Switzerland 2015
R.G. Underwood, *Fundamentals of Hopf Algebras*, Universitext,
DOI 10.1007/978-3-319-18991-8_4

Recalling our collection one parameter Hopf orders (§3.4), we show that there exists a ring of integers S in some extension L of K for which S is a Galois $H(e')$-extension (here we do have the classical Galois action), that is, $H(e')$ is realizable as a Galois group. Locally, a result due to L. Childs states that every one parameter Hopf order $H(i)$ is realizable as a Galois group.

4.1 Quasitriangular Structures

In this section we introduce almost cocommutative bialgebras and quasitriangular bialgebras. We show that a quasitriangular bialgebra determines a solution to the Quantum Yang–Baxter Equation and show how to compute quasitriangular structures for certain two-dimensional bialgebras and Hopf algebras. We show that almost cocommutative Hopf algebras generalize Hopf algebras in which the coinverse has order 2.

∗ ∗ ∗

Let K be a field. Throughout this section, $\otimes = \otimes_K$. Let B be a K-bialgebra and let $B \otimes B$ be the tensor product K-algebra (§1.2). Let $U(B \otimes B)$ denote the group of units in $B \otimes B$ and let $R \in U(B \otimes B)$.

Definition 4.1.1. The pair (B, R) is **almost cocommutative** if the element R satisfies

$$\tau(\Delta_B(b)) = R\Delta_B(b)R^{-1} \tag{4.1}$$

for all $b \in B$.

If the bialgebra B is cocommutative, then the pair $(B, 1 \otimes 1)$ is almost cocommutative. However, if B is commutative and non-cocommutative, then (B, R) cannot be almost cocommutative for any $R \in U(B \otimes B)$ since in this case (4.1) reduces to the condition for cocommutativity.

Write $R = \sum_{i=1}^{n} a_i \otimes b_i \in U(B \otimes B)$. Let

$$R^{12} = \sum_{i=1}^{n} a_i \otimes b_i \otimes 1 \in B^{\otimes^3},$$

$$R^{13} = \sum_{i=1}^{n} a_i \otimes 1 \otimes b_i \in B^{\otimes^3},$$

$$R^{23} = \sum_{i=1}^{n} 1 \otimes a_i \otimes b_i \in B^{\otimes^3}.$$

Definition 4.1.2. The pair (B, R) is **quasitriangular** if (B, R) is almost cocommutative and the following conditions hold:

$$(\Delta_B \otimes I_B)R = R^{13}R^{23} \tag{4.2}$$

$$(I_B \otimes \Delta_B)R = R^{13}R^{12} \tag{4.3}$$

Clearly, if B is cocommutative, then $(B, 1 \otimes 1)$ is quasitriangular. A **quasitriangular structure** is an element $R \in U(B \otimes B)$ so that (B, R) is quasitriangular. Let (B, R) and (B', R') be quasitriangular bialgebras. Then (B, R), (B', R') are **isomorphic as quasitriangular bialgebras**, written $(B, R) \cong (B', R')$, if there exists a bialgebra isomorphism $\phi : B \to B'$ for which $R' = (\phi \otimes \phi)(R)$. Two quasitriangular structures R, R' on a bialgebra B are **equivalent quasitriangular structures** if $(B, R) \cong (B, R')$ as quasitriangular bialgebras.

Example 4.1.3. Suppose that B is a commutative and non-cocommutative bialgebra, for instance, suppose that $B = KG^*$ for G finite non-abelian. Then (B, R) cannot be quasitriangular for any $R \in U(B \otimes B)$; B has no quasitriangular structures.

Example 4.1.4. Let $T = \{1, a\}$ be the monoid with multiplication table

	1	a
1	1	a
a	a	a

Let KT be the monoid bialgebra. We've seen this bialgebra before-it is isomorphic to the Myhill–Nerode bialgebra of Example 2.2.12. As we will show, KT has only the trivial quasitriangular structure $R = 1 \otimes 1$.

Example 4.1.5. Let K be a field of characteristic $\neq 2$, let C_2 be the cyclic group of order 2 generated by g and let KC_2 be the group bialgebra. Then there are exactly two non-equivalent quasitriangular structures on KC_2, namely, $R_0 = 1 \otimes 1$ and

$$R_1 = \frac{1}{2}(1 \otimes 1 + 1 \otimes g + g \otimes 1 - g \otimes g).$$

Example 4.1.6. Let H be the Sweedler Hopf algebra of Example 3.1.5, defined over the field $K = \mathbb{Q}$. For $a \in K$, let

$$R^{(a)} = \frac{1}{2}(1 \otimes 1 + 1 \otimes g + g \otimes 1 - g \otimes g)$$

$$+ \frac{a}{2}(x \otimes x - x \otimes gx + gx \otimes x + gx \otimes gx)$$

Then $R^{(a)}$ is a quasitriangular structure for H. Moreover, there are an infinite number of non-equivalent quasitriangular structures of the form $R^{(a)}$ for H, cf. [Mo93, 10.1.17], [Ra93].

We want to provide some details on how to classify the quasitriangular structures in Examples 4.1.4 and 4.1.5, but first we explain why it is worthwhile to study quasitriangular bialgebras.

Let A be any K-algebra and let $R = \sum_{i=1}^{n} a_i \otimes b_i \in A \otimes A$. As before, let $R^{12} = \sum_{i=1}^{n} a_i \otimes b_i \otimes 1_A$, $R^{13} = \sum_{i=1}^{n} a_i \otimes 1_A \otimes b_i$ and $R^{23} = \sum_{i=1}^{n} 1_A \otimes a_i \otimes b_i$.

Proposition 4.1.7 (Drinfeld [Dr90]). *Suppose (B, R) is a quasitriangular bialgebra. Then*

$$R^{12}R^{13}R^{23} = R^{23}R^{13}R^{12}. \tag{4.4}$$

Proof. One has

$$R^{12}R^{13}R^{23} = R^{12}(\Delta_B \otimes I_B)(R) \quad \text{by (4.2)}$$

$$= (R \otimes 1)\left(\sum_{i=1}^{n} \Delta_B(a_i) \otimes b_i \right)$$

$$= \sum_{i=1}^{n} R\Delta_B(a_i) \otimes b_i$$

$$= \sum_{i=1}^{n} \tau\Delta_B(a_i)R \otimes b_i \quad \text{by (4.1)}$$

$$= \left(\sum_{i=1}^{n} \tau\Delta_B(a_i) \otimes b_i \right)(R \otimes 1)$$

$$= (\tau\Delta_B \otimes I_B)\left(\sum_{i=1}^{n} a_i \otimes b_i \right)(R \otimes 1).$$

Continuing with this calculation, we have

$$(\tau\Delta_B \otimes I_B)\left(\sum_{i=1}^{n} a_i \otimes b_i \right)(R \otimes 1) = (\tau\Delta_B \otimes I_B)(R)R^{12}$$

$$= (\tau \otimes I_B)(\Delta_B \otimes I_B)(R)R^{12}$$

$$= (\tau \otimes I_B)(R^{13}R^{23})R^{12} \quad \text{by (4.2)}$$

$$= R^{23}R^{13}R^{12}.$$

$\square$

Equation (4.4) is known as the **quantum Yang–Baxter equation (QYBE)**, see [Mo93, Chapter 10], [Ni12]. Proposition 4.1.7 says that quasitriangular bialgebras determine solutions to the QYBE. In §4.2 and §4.3 we will show that solutions to the QYBE will yield representations of a certain group called the Braid group. For the present, however, we return to the problem of how to construct quasitriangular structures for the monoid bialgebra KT and the group bialgebra KC_2.

We need the following proposition which shows that every bialgebra isomorphism $\phi : B \to B'$ with B quasitriangular extends to an isomorphism of quasitriangular bialgebras.

Proposition 4.1.8. *Suppose (B, R) is quasitriangular and suppose that $\phi : B \to B'$ is an isomorphism of K-bialgebras. Let $R' = (\phi \otimes \phi)(R)$. Then (B', R') is quasitriangular.*

Proof. Note that $(\phi \otimes \phi)(R^{-1}) = ((\phi \otimes \phi)(R))^{-1}$. Let $b' \in B'$. Then there exists $b \in B$ for which $\phi(b) = b'$. Now

$$
\begin{aligned}
\tau\Delta_{B'}(b') &= \tau\Delta_{B'}(\phi(b)) \\
&= \tau(\phi \otimes \phi)\Delta_B(b) \quad \text{since } \phi \text{ is a bialgebra hom.} \\
&= (\phi \otimes \phi)\tau\Delta_B(b) \\
&= (\phi \otimes \phi)(R\Delta_B(b)R^{-1}) \quad \text{since } (B, R) \text{ is quasitriangular.}
\end{aligned}
$$

Continuing with the calculation, we obtain

$$
\begin{aligned}
(\phi \otimes \phi)(R\Delta_B(b)R^{-1}) &= (\phi \otimes \phi)(R)(\phi \otimes \phi)\Delta_B(b)(\phi \otimes \phi)(R^{-1}) \\
&= (\phi \otimes \phi)(R)\Delta_{B'}(\phi(b))((\phi \otimes \phi)(R))^{-1} \\
&= (\phi \otimes \phi)(R)\Delta_{B'}(b')((\phi \otimes \phi)(R))^{-1} \\
&= R'\Delta_{B'}(b')(R')^{-1}
\end{aligned}
$$

and so, (B, R') is almost cocommutative. We next show that condition (4.2) holds. Indeed,

$$
\begin{aligned}
(\Delta_{B'} \otimes I_{B'})(R') &= (\Delta_{B'} \otimes I_{B'})(\phi \otimes \phi)(R) \\
&= (\Delta_{B'} \otimes I_{B'})\left(\sum_{i=1}^{n} \phi(a_i) \otimes \phi(b_i) \right) \\
&= \sum_{i=1}^{n} \Delta_{B'}(\phi(a_i)) \otimes \phi(b_i) \\
&= \sum_{i=1}^{n} (\phi \otimes \phi)\Delta_B(a_i) \otimes \phi(b_i) \\
&= (\phi \otimes \phi \otimes \phi)\left(\sum_{i=1}^{n} \Delta_B(a_i) \otimes b_i \right) \\
&= (\phi \otimes \phi \otimes \phi)(\Delta_B \otimes I_B)(R).
\end{aligned}
$$

Computing further, one obtains

$$(\phi \otimes \phi \otimes \phi)(\Delta_B \otimes I_B)(R) = (\phi \otimes \phi \otimes \phi)(R^{13}R^{23}) \quad \text{since } (B,R) \text{ is quasitriangular}$$

$$= (\phi \otimes \phi \otimes \phi)\left(\left(\sum_{i=1}^{n} a_i \otimes 1 \otimes b_i\right)\left(\sum_{i=1}^{n} 1 \otimes a_i \otimes b_i\right)\right)$$

$$= \left(\sum_{i=1}^{n} \phi(a_i) \otimes 1 \otimes \phi(b_i)\right)\left(\sum_{i=1}^{n} 1 \otimes \phi(a_i) \otimes \phi(b_i)\right)$$

$$= ((\phi \otimes \phi)(R))^{13}((\phi \otimes \phi)(R))^{23}$$

$$= (R')^{13}(R')^{23},$$

thus (4.2) holds for (B', R'). In a similar manner we get

$$(I_{B'} \otimes \Delta_{B'})(R') = (R')^{13}(R')^{12}$$

Thus (B', R') is quasitriangular. $\square$

To find quasitriangular structures we also employ the following proposition due to Drinfeld [Dr86]. Let $s_1 : K \otimes B \to B$, $s_2 : B \otimes K \to B$ be the maps defined as $r \otimes b \mapsto rb$, $b \otimes r \mapsto rb$, respectively.

Proposition 4.1.9 (Drinfeld). *Suppose* (B, R) *is quasitriangular. Then*

(i) $s_1(\epsilon_B \otimes I_B)(R) = 1$,
(ii) $s_2(I_B \otimes \epsilon_B)(R) = 1$.

Proof. Put $I = I_B, \epsilon = \epsilon_B, \Delta = \Delta_B$. We prove (i). First observe that

$$(s_1 \otimes I)(\epsilon \otimes I \otimes I)(\Delta \otimes I)(R) = (s_1 \otimes I)(\epsilon \otimes I \otimes I)\left(\sum_{i=1}^{n} \Delta_B(a_i) \otimes b_i\right)$$

$$= (s_1 \otimes I)\left(\sum_{i=1}^{n}(\epsilon \otimes I)\Delta(a_i) \otimes b_i\right)$$

$$= \sum_{i=1}^{n} s_1(\epsilon \otimes I)\Delta(a_i) \otimes b_i$$

$$= \sum_{i=1}^{n} a_i \otimes b_i$$

$$= R.$$

And so,

$$R = (s_1 \otimes I)(\epsilon \otimes I \otimes I)(\Delta \otimes I)(R)$$

$$= (s_1 \otimes I)(\epsilon \otimes I \otimes I)(R^{13}R^{23}) \quad \text{by (4.2)}$$

$$= (s_1 \otimes I)(\epsilon \otimes I \otimes I)(R^{13})(s_1 \otimes I)(\epsilon \otimes I \otimes I)(R^{23})$$

$$= (s_1 \otimes I)(\epsilon \otimes I \otimes I)\left(\sum_{i=1}^{n} a_i \otimes 1 \otimes b_i \right)(s_1 \otimes I)(\epsilon \otimes I \otimes I)\left(\sum_{i=1}^{n} 1 \otimes a_i \otimes b_i \right)$$

$$= \left(\sum_{i=1}^{n} \epsilon(a_i) 1 \otimes b_i \right)\left(\sum_{i=1}^{n} a_i \otimes b_i \right)$$

$$= \left(\sum_{i=1}^{n} 1 \otimes \epsilon(a_i) b_i \right) R.$$

Thus

$$1 \otimes \sum_{i=1}^{n} \epsilon(a_i) b_i = 1 \otimes 1$$

and consequently,

$$1 = s_1\left(\sum_{i=1}^{n} \epsilon(a_i) \otimes b_i \right) = s_1(\epsilon \otimes I)(R).$$

A similar argument is used to prove (ii). $\square$

Now we can show that the monoid bialgebra $B = KT$ of Example 4.1.4 has only the trivial quasitriangular structure. Observe that the linear dual KT^* is a K-bialgebra on the basis $\{e_1, e_a\}$ with $e_x(y) = \delta_{x,y}$, The algebra structure on KT^* is given by $e_x e_y = \delta_{x,y} e_x$. By Proposition 1.3.10, comultiplication on KT^* is defined by

$$\Delta_{KT^*}(e_1) = e_1 \otimes e_1$$

$$\Delta_{KT^*}(e_a) = e_1 \otimes e_a + e_a \otimes e_1 + e_a \otimes e_a.$$

The counit map is defined by

$$\epsilon_{KT^*}(e_1) = 1, \quad \epsilon_{KT^*}(e_a) = 0.$$

There is a bialgebra isomorphism $\phi : KT \to KT^*$ defined as $\phi(1) = e_1 + e_a$, $\phi(a) = e_1$.

Proposition 4.1.10. *Let KT be the K-bialgebra of Example 4.1.4. Then there is exactly one quasitriangular structure on KT, namely, $R = 1 \otimes 1$.*

Proof. Certainly, $1 \otimes 1$ is a quasitriangular structure for KT. We claim that $1 \otimes 1$ is the only quasitriangular structure.

If (KT, R) is quasitriangular, then (KT^*, R'), $R' = (\phi \otimes \phi)(R)$, is quasitriangular by Proposition 4.1.8. So, we first compute all of the quasitriangular structures of KT^*. To this end suppose that (KT^*, R') is quasitriangular for some element $R' \in KT^* \otimes KT^*$. Since

$$KT^* \otimes KT^* = K(e_1 \otimes e_1) \oplus K(e_1 \otimes e_a) \oplus K(e_a \otimes e_1) \oplus K(e_a \otimes e_a),$$

$$R' = w(e_1 \otimes e_1) + x(e_1 \otimes e_a) + y(e_a \otimes e_1) + z(e_a \otimes e_a)$$

for $w, x, y, z \in K$. Put $1 = 1_{KT^*}$, $I = I_{KT^*}$, $\epsilon = \epsilon_{KT^*}$, $\Delta = \Delta_{KT^*}$. By Proposition 4.1.9(i),

$$
\begin{aligned}
1 &= e_1 + e_a \\
&= s_1(\epsilon \otimes I)(w(e_1 \otimes e_1) + x(e_1 \otimes e_a) + y(e_a \otimes e_1) + z(e_a \otimes e_a)) \\
&= we_1 + xe_a
\end{aligned}
$$

and so, $w = x = 1$. From Proposition 4.1.9(ii), one also has $y = 1$. Thus

$$R' = e_1 \otimes e_1 + e_1 \otimes e_a + e_a \otimes e_1 + z(e_a \otimes e_a)$$

for $z \in K$. We want to find all values of z for which (KT^*, R') is quasitriangular, necessarily, we require that

$$(\Delta \otimes I)(R') = (R')^{13}(R')^{23}.$$

Now,

$$
\begin{aligned}
(\Delta \otimes I)(R') &= (\Delta \otimes I)(e_1 \otimes e_1 + e_1 \otimes e_a + e_a \otimes e_1 + z(e_a \otimes e_a)) \\
&= (e_1 \otimes e_1) \otimes e_1 + (e_1 \otimes e_1) \otimes e_a + (e_1 \otimes e_a + e_a \otimes e_1 + e_a \otimes e_a) \otimes e_1 \\
&\quad + z((e_1 \otimes e_a + e_a \otimes e_1 + e_a \otimes e_a) \otimes e_a) \\
&= e_1 \otimes e_1 \otimes e_1 + e_1 \otimes e_1 \otimes e_a + e_1 \otimes e_a \otimes e_1 + e_a \otimes e_1 \otimes e_1 \\
&\quad + e_a \otimes e_a \otimes e_1 + z(e_1 \otimes e_a \otimes e_a) + z(e_a \otimes e_1 \otimes e_a) \\
&\quad + z(e_a \otimes e_a \otimes e_a).
\end{aligned}
\tag{4.5}
$$

Moreover,

$$
\begin{aligned}
(R')^{13}(R')^{23} &= (e_1 \otimes (e_1 + e_a) \otimes e_1 + e_1 \otimes (e_1 + e_a) \otimes e_a + e_a \otimes (e_1 + e_a) \otimes e_1 \\
&\quad + z(e_a \otimes (e_1 + e_a) \otimes e_a))((e_1 + e_a) \otimes e_1 \otimes e_1 + (e_1 + e_a) \otimes e_1 \otimes e_a \\
&\quad + (e_1 + e_a) \otimes e_a \otimes e_1 + z((e_1 + e_a) \otimes e_a \otimes e_a)) \\
&= (e_1 \otimes e_1 \otimes e_1 + e_1 \otimes e_a \otimes e_1 + e_1 \otimes e_1 \otimes e_a + e_1 \otimes e_a \otimes e_a \\
&\quad + e_a \otimes e_1 \otimes e_1 + e_a \otimes e_a \otimes e_1 + z(e_a \otimes e_1 \otimes e_a) \\
&\quad + z(e_a \otimes \otimes e_a \otimes e_a))(e_1 \otimes e_1 \otimes e_1 + e_a \otimes e_1 \otimes e_1 + e_1 \otimes e_1 \otimes e_a \\
&\quad + e_a \otimes e_a \otimes e_1 + e_1 \otimes e_a \otimes e_1 + e_a \otimes e_a \otimes e_1 + z(e_1 \otimes e_a \otimes e_a) \\
&\quad + z(e_a \otimes \otimes e_a \otimes e_a))
\end{aligned}
$$

$$= e_1 \otimes e_1 \otimes e_1 + e_1 \otimes e_a \otimes e_1 + e_1 \otimes e_1 \otimes e_a + z(e_1 \otimes e_a \otimes e_a)$$
$$+ e_a \otimes e_1 \otimes e_1 + e_a \otimes e_a \otimes e_1 + z(e_a \otimes e_1 \otimes e_a)$$
$$+ z^2(e_a \otimes e_a \otimes e_a). \tag{4.6}$$

Thus if (KT^*, R') is quasitriangular, then from (4.5) and (4.6)

$$e_1 \otimes e_1 \otimes e_1 + e_1 \otimes e_1 \otimes e_a + e_1 \otimes e_a \otimes e_1 + e_a \otimes e_1 \otimes e_1$$
$$+ e_a \otimes e_a \otimes e_1 + z(e_1 \otimes e_a \otimes e_a) + z(e_a \otimes e_1 \otimes e_a)$$
$$+ z(e_a \otimes e_a \otimes e_a)$$

$$= e_1 \otimes e_1 \otimes e_1 + e_1 \otimes e_a \otimes e_1 + e_1 \otimes e_1 \otimes e_a + z(e_1 \otimes e_a \otimes e_a)$$
$$+ e_a \otimes e_1 \otimes e_1 + e_a \otimes e_a \otimes e_1 + z(e_a \otimes e_1 \otimes e_a)$$
$$+ z^2(e_a \otimes e_a \otimes e_a),$$

and so, $z^2 = z$. Thus either $z = 0$ or $z = 1$. If $z = 0$, then R' is not a unit in $KT^* \otimes KT^*$. Thus

$$R' = e_1 \otimes e_1 + e_1 \otimes e_a + e_a \otimes e_1 + e_a \otimes e_a = 1 \otimes 1$$

is the only quasitriangular structure for KT^*. Consequently, if (KT, R) is quasitriangular, then

$$(\phi \otimes \phi)(R) = 1 \otimes 1.$$

It follows that $R = 1 \otimes 1$.   □

Let H be a K-Hopf algebra and let $R = \sum_{i=1}^{n} a_i \otimes b_i \in U(H \otimes H)$.

Definition 4.1.11. The pair (H, R) is a **quasitriangular Hopf algebra** if (H, R) is quasitriangular as a K-bialgebra and the coinverse map σ_H is a bijection.

Recall this means that the following conditions hold:

$$(\Delta_H \otimes I_H)R = R^{13}R^{23}, \tag{4.7}$$

$$(I_H \otimes \Delta_H)R = R^{13}R^{12}. \tag{4.8}$$

We note that if H is a finite dimensional Hopf algebra, then its coinverse map σ_H is necessarily bijective (this follows from Proposition 3.2.21, see [Mo93, (2.1.3)]). Thus Hopf algebras that were quasitriangular as bialgebras, yet did not have bijective coinverses, would not include finite dimensional Hopf algebras.

A **quasitriangular structure** is an element $R \in U(H \otimes H)$ so that (H, R) is quasitriangular. Let (H, R) and (H', R') be quasitriangular Hopf algebras. Then (H, R),

(H', R') are **isomorphic as quasitriangular Hopf algebras** if there exists a Hopf algebra isomorphism $\phi : H \to H'$ for which $R' = (\phi \otimes \phi)(R)$. Two quasitriangular structures R, R' on H are **equivalent quasitriangular structures** if $(H, R) \cong (H, R')$ as quasitriangular Hopf algebras.

Our goal is to compute all of the quasitriangular structures on the Hopf algebra KC_2 of Example 4.1.5. First, we prove a proposition.

Proposition 4.1.12. *Suppose (H, R) is quasitriangular. Then $(\sigma_H \otimes I_H)(R) = R^{-1}$.*

Proof. One has

$$
R(\sigma_H \otimes I_H)(R) = \left(\sum_{i=1}^{n} a_i \otimes b_i \right) (\sigma_H \otimes I) \left(\sum_{j=1}^{n} a_j \otimes b_j \right)
$$

$$
= \left(\sum_{i=1}^{n} a_i \otimes b_i \right) \left(\sum_{j=1}^{n} a_i \sigma_H(a_j) \otimes b_i b_j \right)
$$

$$
= \sum_{i=1}^{n} \sum_{j=1}^{n} a_i \sigma_H(a_j) \otimes b_i b_j
$$

$$
= (m_H \otimes I_H)(I_H \otimes \sigma_H \otimes I_H) \left(\sum_{i=1}^{n} \sum_{j=1}^{n} a_i \otimes a_j \otimes b_i b_j \right)
$$

$$
= (m_H \otimes I_H)(I_H \otimes \sigma_H \otimes I_H)(R^{13} R^{23}).
$$

Continuing with the calculation,

$$
(m_H \otimes I_H)(I_H \otimes \sigma_H \otimes I_H)(R^{13} R^{23})
$$

$$
= (m_H \otimes I_H)(I_H \otimes \sigma_H \otimes I_H)(\Delta_H \otimes I_H)(R) \quad \text{by (4.7)}
$$

$$
= \sum_{i=1}^{n} \epsilon_H(a_i) 1_H \otimes b_i \quad \text{by the coinverse property}
$$

$$
= 1_H \otimes 1_H \quad \text{by Proposition 4.1.9(i).}
$$

Thus $(\sigma_H \otimes I_H)(R) = R^{-1}$. $\square$

Proposition 4.1.13. *Let KC_2 be the Hopf algebra of Example 4.1.5. There are exactly two quasitriangular structures on KC_2, namely, $R_0 = 1 \otimes 1$ and*

$$
R_1 = \frac{1+g}{2} \otimes 1 + \frac{1-g}{2} \otimes g.
$$

Proof. Certainly, $R_0 = 1 \otimes 1$ is a quasitriangular structure for KC_2.

Let $\{p_1, p_g\}$ be the basis for KC_2^* dual to the basis $\{1, g\}$ for KC_2. The dual KC_2^* is a Hopf algebra with comultiplication defined by

$$\Delta_{KC_2^*}(p_1) = p_1 \otimes p_1 + p_g \otimes p_g,$$
$$\Delta_{KC_2^*}(p_g) = p_1 \otimes p_g + p_g \otimes p_1,$$

counit map given by $\epsilon_{KC_2^*}(p_1) = 1$, $\epsilon_{KC_2^*}(p_g) = 0$, and coinverse defined by $\sigma_{KC_2^*}(p_1) = p_1$, $\sigma_{KC_2^*}(p_g) = p_g$.

Note that $\phi : KC_2 \to KC_2^*$ defined as $\phi(1) = p_1 + p_g$, $\phi(g) = p_1 - p_g$, is an isomorphism of Hopf algebras. If (KC_2, R) is quasitriangular, then (KC_2^*, R'), $R' = (\phi \otimes \phi)(R)$, is quasitriangular by Proposition 4.1.8. So, we first compute all of the quasitriangular structures of KC_2^*. To this end suppose that (KC_2^*, R') is quasitriangular for some element $R' \in U(KC_2^* \otimes KC_2^*)$. Then

$$R' = w(p_1 \otimes p_1) + x(p_1 \otimes p_g) + y(p_g \otimes p_1) + z(p_g \otimes p_g)$$

for $w, x, y, z \in K$. By Proposition 4.1.9(i), $1 = wp_1 + xp_g$, and so, $w = x = 1$. From Proposition 4.1.9(ii), one also has $y = 1$. Thus

$$R' = p_1 \otimes p_1 + p_1 \otimes p_g + p_g \otimes p_1 + z(p_g \otimes p_g),$$

for $z \in K$. Now by Proposition 4.1.12

$$(\sigma_{KC_2^*} \otimes I_{KC_2^*})(R') = R' = (R')^{-1},$$

and so, $z^2 = 1$. Thus, either $z = 1$, which yields the quasitriangular structure $1 \otimes 1$, or $z = -1$, which yields the unit

$$R' = p_1 \otimes p_1 + p_1 \otimes p_g + p_g \otimes p_1 - (p_g \otimes p_g).$$

As one can verify, (KC_2^*, R') is quasitriangular. Consequently,

$$\phi^{-1}(R') = R_1 = \frac{1+g}{2} \otimes 1 + \frac{1-g}{2} \otimes g$$

is a quasitriangular structure for KC_2. $\qquad\square$

Let H be a K-Hopf algebra. As we have seen, if H is cocommutative, then the pair $(H, 1 \otimes 1)$ is almost cocommutative. By Proposition 3.1.9, cocommutativity of H implies that

$$\sigma_H^2(h) = h$$

for all $h \in H$. The generalization of this property for almost cocommutative (H, R) is due to Drinfeld [Dr90].

Proposition 4.1.14 (Drinfeld). *Let (H, R) be almost cocommutative with $R = \sum_{i=1}^{n} a_i \otimes b_i \in U(H \otimes H)$. Let $u = \sum_{i=1}^{n} \sigma_H(b_i) a_i$. Then $u \in U(H)$ and*

$$\sigma_H^2(h) = uhu^{-1},$$

for all $h \in H$.

Proof. Let $h \in H$. Since (H, R) is almost cocommutative,

$$R\Delta_H(h) = \left(\sum_{i=1}^{n} a_i \otimes b_i \right) \left(\sum_{(h)} h_{(1)} \otimes h_{(2)} \right)$$

$$= \sum_{(h)} \sum_{i=1}^{n} a_i h_{(1)} \otimes b_i h_{(2)}$$

is equal to

$$\tau\Delta_H(h)R = \tau\left(\sum_{(h)} h_{(1)} \otimes h_{(2)} \right) \left(\sum_{i=1}^{n} a_i \otimes b_i \right)$$

$$= \tau\left(\sum_{(h)} h_{(2)} \otimes h_{(1)} \right) \left(\sum_{i=1}^{n} a_i \otimes b_i \right)$$

$$= \sum_{(h)} \sum_{i=1}^{n} h_{(2)} a_i \otimes h_{(1)} b_i.$$

Consequently,

$$\sum_{(h)} \sum_{i=1}^{n} a_i h_{(1)} \otimes b_i h_{(2)} \otimes h_{(3)} = \sum_{(h)} \sum_{i=1}^{n} h_{(2)} a_i \otimes h_{(1)} b_i \otimes h_{(3)}, \qquad (4.9)$$

where $h_{(3)}$ is so that

$$\Delta_H(h) = (I_H \otimes \Delta_H)\Delta_H(h) = \sum_{(h)} h_{(1)} \otimes h_{(2)} \otimes h_{(3)}.$$

From (4.9), one obtains

$$\sum_{(h)} \sum_{i=1}^{n} \sigma_H(\sigma_H(h_{(3)}))\sigma_H(b_i h_{(2)}) a_i h_{(1)}$$

$$= \sum_{(h)} \sum_{i=1}^{n} \sigma_H(\sigma_H(h_{(3)}))\sigma_H(h_{(1)} b_i) h_{(2)} a_i. \qquad (4.10)$$

Now,

$$\sum_{(h)}\sum_{i=1}^{n}\sigma_H(\sigma_H(h_{(3)}))\sigma_H(b_ih_{(2)})a_ih_{(1)}$$

$$=\sum_{(h)}\sum_{i=1}^{n}\sigma_H(\sigma_H(h_{(3)}))\sigma_H(h_{(2)})\sigma_H(b_i)a_ih_{(1)}, \quad \text{by Proposition 3.1.8(i)}$$

$$=\sum_{(h)}\sum_{i=1}^{n}\sigma_H(h_{(2)}\sigma_H(h_{(3)}))\sigma_H(b_i)a_ih_{(1)}, \quad \text{by Proposition 3.1.8(i)}$$

$$=\sum_{(h)}\sum_{i=1}^{n}\sigma_H(\epsilon_H(h_{(2)})1_H)\sigma_H(b_i)a_ih_{(1)} \quad \text{by the coinverse property}$$

$$=\sum_{i=1}^{n}\sigma_H(b_i)a_ih \quad \text{by the counit property}$$

$$= uh.$$

On the other hand,

$$\sum_{(h)}\sum_{i=1}^{n}\sigma_H(\sigma_H(h_{(3)}))\sigma_H(h_{(1)}b_i)h_{(2)}a_i$$

$$=\sum_{(h)}\sum_{i=1}^{n}\sigma_H(\sigma_H(h_{(3)}))\sigma_H(b_i)\sigma_H(h_{(1)})h_{(2)}a_i, \quad \text{by Proposition 3.1.8(i)}$$

$$=\sum_{(h)}\sum_{i=1}^{n}\sigma_H(\sigma_H(h_{(2)}))\sigma_H(b_i)\epsilon_H(h_{(1)})a_i \quad \text{by the coinverse property}$$

$$=\sum_{i=1}^{n}\sigma_H(\sigma_H(h))\sigma_H(b_i)a_i \quad \text{by Proposition 3.1.8(ii) and the counit property}$$

$$= \sigma_H^2(h)u.$$

And so, (4.10) yields

$$uh = \sigma_H^2(h)u, \tag{4.11}$$

for all $h \in H$.

It remains to show that u is a unit in H. To this end, let $R^{-1} = \sum_{i=1}^{m} c_j \otimes d_j$ and let $v = \sum_{j=1}^{m} \sigma_H^{-1}(d_j)c_j$ (recall that σ_H is a bijection). Now,

$$uv = u \sum_{j=1}^{m} \sigma_H^{-1}(d_j) c_j$$

$$= \sum_{j=1}^{m} (u \sigma_H^{-1}(d_j)) c_j$$

$$= \sum_{j=1}^{m} (\sigma_H^2(\sigma_H^{-1}(d_j) u) c_j \quad \text{by (4.11)}$$

$$= \sum_{j=1}^{m} \sigma_H(d_j) u c_j$$

$$= \sum_{j=1}^{m} \sigma_H(d_j) \sum_{i=1}^{n} \sigma_H(b_i) a_i c_j$$

$$= \sum_{j=1}^{m} \sum_{i=1}^{n} \sigma_H(b_i d_j) a_i c_j \quad \text{by Proposition 3.1.8(i)}.$$

Moreover,

$$\sum_{j=1}^{m} \sum_{i=1}^{n} \sigma_H(b_i d_j) a_i c_j = m_H(\sigma_H \otimes I_H) \left(\sum_{i=1}^{n} \sum_{j=1}^{m} b_i d_j \otimes a_i c_j \right)$$

$$= m_H(\sigma_H \otimes I_H)(1_H \otimes 1_H) \quad \text{since } RR^{-1} = 1_H \otimes 1_H.$$

$$= \sigma_H(1_H)$$

$$= 1_H \quad \text{by Proposition 3.1.8(ii)}.$$

Consequently, $uv = 1_H$. Now,

$$1 = uv = \sigma_H^2(v) u,$$

by (4.11), and so $v = \sigma_H^2(v)$. It follows that v is so that

$$uv = 1_H = vu$$

and so u is a unit of H. Consequently, (4.11) is now

$$\sigma_H^2(h) u = uhu^{-1},$$

for all $h \in H$. $\square$

4.2 The Braid Group

In this section we define a certain infinite group called the braid group on three strands. The braid group is generated by two fundamental braids B_1, B_2 which satisfy the braid relation $B_1 B_2 B_1 = B_2 B_1 B_2$. We will use the braid group in the next section.

$$* \quad * \quad *$$

Let $\mathbb{R}^3$ denote Euclidean 3-space together with the familiar xyz-coordinate system in the standard orientation. Let ℓ denote the line containing the point $(0, 1, 0)$ that is perpendicular to the xy-plane. Let C_1 be a smooth curve (a path) in $\mathbb{R}^3$ starting at the point $(0, 0, 1)$ on the z-axis and ending at one of the three points $(0, 1, 1)$, $(0, 1, 2)$, $(0, 1, 3)$ on ℓ; let C_2 be a smooth curve in $\mathbb{R}^3$ starting at the point $(0, 0, 2)$ on the z-axis and ending at one of the three points $(0, 1, 1)$, $(0, 1, 2)$, $(0, 1, 3)$ on ℓ that is not the terminal point of C_1. We assume that C_1 and C_2 are disjoint. Let C_3 be a smooth curve in $\mathbb{R}^3$ starting at the point $(0, 0, 3)$ on the z-axis and ending at one of the three points $(0, 1, 1)$, $(0, 1, 2)$, $(0, 1, 3)$ on ℓ that is not the terminal point of C_1 or C_2. We require that the paths C_1, C_2, and C_3 are pairwise disjoint. The resulting collection of paths is a **braid on three strands**, or more simply, a **braid**. See Figure 4.1 for an example of a braid.

Let B, B' be braids with paths C_1, C_2, C_3 and C_1', C_2', C_3', respectively. Then B, B' are **equivalent** if there exist *path homotopies*

$$F_i : I \times I \to \mathbb{R}^3,$$

$I = [0, 1]$, $i = 1, 2, 3$, for which $F_i(x, 0) = C_i$, $F_i(x, 1) = C_i'$ and for which $i \neq j$ implies that $F_i(x, s) \neq F_j(y, t)$ for all $x, y, s, t \in I$. This means that for $i \neq j$, the path C_i is deformed to the corresponding path C_i' in such a way that no intermediate path in the deformation of C_i intersects any intermediate path in the deformation of

Fig. 4.1 A braid B in $\mathbb{R}^3$

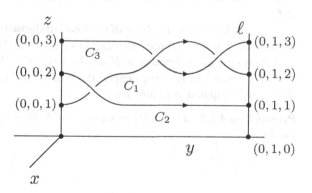

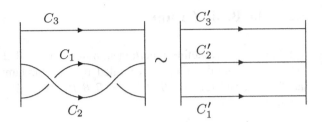

Fig. 4.2 Equivalent braids in $\mathbb{R}^3$

C_j to C_j'. In other words two braids B, B' are equivalent if they are essentially the same in terms of the way the paths are interwoven. In Figure 4.2 below, we given an example of two braids that are equivalent.

Braid equivalence is an equivalence relation on the collection of braids in $\mathbb{R}^3$. Consequently, the collection of braids in $\mathbb{R}^3$ can be partitioned into a set $\mathcal{B}$ of equivalence classes of braids. We let $[B] \in \mathcal{B}$ denote the equivalence class represented by the braid B.

We define a binary operation on $\mathcal{B}$ as follows. Let $[B], [B'] \in \mathcal{B}$. Place B next to B' so that the right axis (ℓ) of B coincides with the left axis (z-axis) of B' and then erase the middle axis. The result is a braid which we denote as BB' (see Figure 4.4 for an illustration); the corresponding equivalence class is $[BB']$; the binary operation on $\mathcal{B}$ is defined as

$$[B][B'] = [BB'].$$

This is indeed a well-defined binary operation on $\mathcal{B}$: if $A \in [B]$, $A' \in [B']$, then $[A][A'] = [AA'] = [BB']$. The braid product on $\mathcal{B}$ is easily seen to be associative: for $[B], [B'], [B''] \in \mathcal{B}$, one has

$$[B]([B'][B'']) = ([B][B'])[B''].$$

In what follows, we simplify matters and identify the braid B with its braid class $[B]$ in $\mathcal{B}$.

The simplest braids are those of the **four fundamental braids**: B_1, B_2, B_{-1}, B_{-2}, given in the (simplified) braid diagrams in Figure 4.3, below.

Amazingly, every braid B in $\mathcal{B}$ can be written as the product of finite number of fundamental braids. For example, the braid B of Figure 4.1 is the product $B = B_1 B_{-2}^2$, as computed in Figure 4.4.

Proposition 4.2.1. *The collection $\mathcal{B}$ of braids in $\mathbb{R}^3$ is a group under the braid product.*

Fig. 4.3 The four
fundamental braids

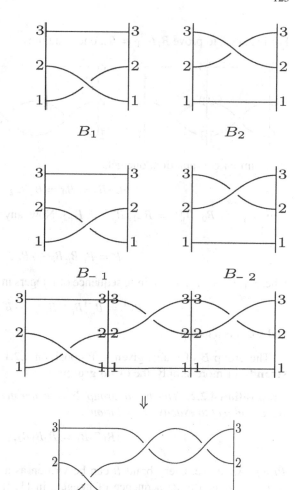

Fig. 4.4 The braid product
$B = B_1 B_{-2}^2$

Proof. We have already seen that the braid product is associative. Let B_0 denote the
braid

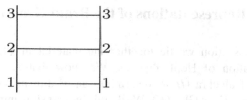

and let B be any braid. Then $B_0 B = B = B B_0$ as one can easily check, and so
we choose B_0 as the identity element. We need to show that every braid admits an
inverse under the braid product. Clearly $B_0^2 = B_0$. Moreover,

$$B_{-1} B_1 = B_0 = B_1 B_{-1}.$$

For instance, to prove $B_1 B_{-1} = B_0$, one computes

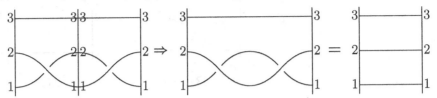

By a similar computation, one gets

$$B_{-2} B_2 = B_0 = B_2 B_{-2}.$$

Thus $B_0^{-1} = B_0$, $B_1^{-1} = B_{-1}$, $B_2^{-1} = B_{-2}$. Now, any braid B can be written as the word

$$B = B_{i_1} B_{i_2} B_{i_3} \cdots B_{i_k},$$

where $i_1, i_2, i_3, \ldots, i_k$ is a finite sequence of integers in $\{-2, -1, 1, 2\}$. It follows that

$$B^{-1} = B_{i_k}^{-1} B_{i_{k-1}}^{-1} B_{i_{k-2}}^{-1} \cdots B_{i_1}^{-1},$$

and so, $\mathcal{B}$ is a group. $\square$

The group $\mathcal{B}$ of braids given in Proposition 4.2.1 is the **braid group on three strands**, or more simply, the **braid group**.

Proposition 4.2.2. *The braid group $\mathcal{B}$ is generated by the fundamental braids B_1, B_2 subject to exactly one relation*

$$B_1 B_2 B_1 = B_2 B_1 B_2. \tag{4.12}$$

Proof. As above, every braid B can be written as a product $B_{i_1}^{n_1} B_{i_2}^{n_2} \cdots B_{i_k}^{n_k}$, where $i_1, i_2, \ldots, i_k$ is a finite sequence of integers in $\{1, 2\}$ and $n_1, n_2, \ldots, n_k \in \mathbb{Z}$. The relation (4.12) follows from the structure of the fundamental braids B_1 and B_2. Any relation amongst the braids B_1 and B_2 reduces to the defining relation (4.12). $\square$

4.3 Representations of the Braid Group

In this section we tie together the material in §4.1, §4.2 and present our first application of Hopf algebras. We show that an n-dimensional quasitriangular K-Hopf algebra (H, R) determines an n^3-dimensional representation of the braid group $\rho : \mathcal{B} \to \mathrm{GL}_{n^3}(K)$. We do this by first showing that the solution of the QYBE (§4.1) yields a braid relation which in turn will determine a group homomorphism $\rho : \mathcal{B} \to \mathrm{GL}_{n^3}(K)$. We give some specific examples of representations determined by the quasitriangular structures of the Hopf algebra KC_2.

$*$ $*$ $*$

Let (H, R) be an n-dimensional quasitriangular Hopf algebra over K and let $\{c_1, c_2, \ldots, c_n\}$ be a K-basis for H. Then $\{c_i \otimes c_j \otimes c_k\}$, $1 \leq i, j, k \leq n$, is a K-basis for the n^3-dimensional tensor product algebra $H \otimes H \otimes H = H^{\otimes^3}$. The matrices in $GL_{n^3}(K)$ correspond to the collection of invertible linear transformations $H^{\otimes^3} \to H^{\otimes^3}$. Some of the matrices in $GL_{n^3}(K)$ arise from the elements $R^{12} = \sum_i a_i \otimes b_i \otimes 1$, $R^{13} = \sum_i a_i \otimes \otimes 1 \otimes b_i$, $R^{23} = \sum_i 1 \otimes a_i \otimes b_i$ constructed from the element $R = \sum_i a_i \otimes b_i \in U(H \otimes H)$. Here is how this happens. For each pair $ij = 12, 13, 23$, let

$$R^{ij} : H^{\otimes^3} \to H^{\otimes^3},$$

be the map defined by left multiplication by R^{ij}. Also, let μ_{ij} be the transposition maps:

$$\mu_{12} : H^{\otimes^3} \to H^{\otimes^3}, \quad x \otimes y \otimes z \mapsto y \otimes x \otimes z,$$

$$\mu_{13} : H^{\otimes^3} \to H^{\otimes^3}, \quad x \otimes y \otimes z \mapsto z \otimes y \otimes x,$$

$$\mu_{23} : H^{\otimes^3} \to H^{\otimes^3}, \quad x \otimes y \otimes z \mapsto x \otimes z \otimes y.$$

Now, define $R_{ij} : H^{\otimes^3} \to H^{\otimes^3}$ to be the composition of maps $R_{ij} = \mu_{ij} R^{ij}$. Note that R_{12} and R_{23} are invertible K-linear transformations of $H^{\otimes^3}$ that (with respect to the K-basis $\{c_i \otimes c_j \otimes c_k\}$) correspond to matrices in $GL_{n^3}(K)$.

Proposition 4.3.1. *Let K be a field and let (H, R) be a quasitriangular Hopf algebra of dimension n over K. Then the matrices R_{12}, R_{23} in $GL_{n^3}(K)$ satisfy*

$$R_{12} R_{23} R_{12} = R_{23} R_{12} R_{23}.$$

Proof. For $x \otimes y \otimes z \in H^{\otimes^3}$, one has

$$(\mu_{13}\mu_{23}R_{23}\mu_{13}\mu_{12})(x \otimes y \otimes z) = (\mu_{13}\mu_{23}R_{23}\mu_{13})(y \otimes x \otimes z)$$

$$= (\mu_{13}\mu_{23}R_{23})(z \otimes x \otimes y)$$

$$= (\mu_{13}\mu_{23})\left(\sum_i z \otimes b_i y \otimes a_i x\right)$$

$$= \mu_{13}\left(\sum_i z \otimes a_i x \otimes b_i y\right)$$

$$= \sum_i b_i y \otimes a_i x \otimes z$$

$$= R_{12}(x \otimes y \otimes z),$$

thus

$$\mu_{13}\mu_{23}R_{23}\mu_{13}\mu_{12} = R_{12}. \tag{4.13}$$

Moreover,

$$(\mu_{12}R_{13}\mu_{12})(x \otimes y \otimes z) = (\mu_{12}R_{13})(y \otimes x \otimes z)$$

$$= \mu_{12}\left(\sum_i b_i z \otimes x \otimes a_i y \right)$$

$$= \sum_i x \otimes b_i z \otimes a_i y$$

$$= R_{23}(x \otimes y \otimes z),$$

and so,

$$\mu_{12}R_{13}\mu_{12} = R_{23}. \tag{4.14}$$

Consequently,

$$R_{12}R_{23}R_{12} = (\mu_{13}\mu_{23}R_{23}\mu_{13}\mu_{12})(\mu_{12}R_{13}\mu_{12})R_{12} \quad \text{by (4.13), (4.14)}$$

$$= \mu_{13}(\mu_{23}R_{23})(\mu_{13}\mu_{12}\mu_{12}R_{13})\mu_{12}R_{12}$$

$$= \mu_{13}(R^{23}R^{13}R^{12})$$

$$= \mu_{13}(R^{12}R^{13}R^{23}) \quad \text{by Proposition 4.1.7.} \tag{4.15}$$

On the other hand,

$$(\mu_{13}\mu_{12}R_{12}\mu_{13}\mu_{23})(x \otimes y \otimes z) = (\mu_{13}\mu_{12}R_{12}\mu_{13})(x \otimes z \otimes y)$$

$$= (\mu_{13}\mu_{12}R_{12})(y \otimes z \otimes x)$$

$$= (\mu_{13}\mu_{12})\left(\sum_i b_i z \otimes a_i y \otimes x \right)$$

$$= \mu_{13}\left(\sum_i a_i y \otimes b_i z \otimes x \right)$$

$$= \sum_i x \otimes b_i z \otimes a_i y$$

$$= R_{23}(x \otimes y \otimes z),$$

and so,

$$\mu_{13}\mu_{12}R_{12}\mu_{13}\mu_{23} = R_{23}. \tag{4.16}$$

Moreover,

$$(\mu_{23}R_{13}\mu_{23})(x \otimes y \otimes z) = (\mu_{23}R_{13})(x \otimes z \otimes y)$$

$$= \mu_{23}\left(\sum_i b_i y \otimes z \otimes a_i x\right)$$

$$= \sum_i b_i y \otimes a_i x \otimes z$$

$$= R_{12}(x \otimes y \otimes z),$$

and so,

$$\mu_{23}R_{13}\mu_{23} = R_{12}. \tag{4.17}$$

Consequently,

$$R_{23}R_{12}R_{23} = (\mu_{13}\mu_{12}R_{12}\mu_{13}\mu_{23})(\mu_{23}R_{13}\mu_{23})R_{23} \quad \text{by (4.16), (4.17)}$$

$$= \mu_{13}R^{12}R^{13}R^{23}$$

$$= R_{12}R_{23}R_{12} \quad \text{by (4.15).}$$

$$\square$$

We now construct our representation of $\mathcal{B}$.

Proposition 4.3.2. *Let (H, R) be a quasitriangular Hopf algebra of dimension n over K. Let $\mathcal{B}$ denote the braid group. Then there exists an n^3-dimensional linear representation*

$$\rho : \mathcal{B} \to GL_{n^3}(K)$$

defined by:

$$\rho(B_1) = R_{12}, \ \rho(B_2) = R_{23}, \ \rho(B_{-1}) = R_{12}^{-1}, \ \rho(B_{-2}) = R_{23}^{-1},$$

and for a braid $B = B_{i_1}B_{i_2}\cdots B_{i_k}$, where $i_1, i_2, \ldots, i_k$ is a finite sequence in $\{-2, -1, 1, 2\}$,

$$\rho(B) = \rho(B_{i_1}B_{i_2}\cdots B_{i_k})$$

$$= \rho(B_{i_1})\rho(B_{i_2})\cdots\rho(B_{i_k}).$$

Proof. We only need to check that ρ respects the braid relation (4.12) of Proposition 4.2.2: $B_1 B_2 B_1 = B_2 B_1 B_2$, but this is easy since

$$
\begin{aligned}
\rho(B_1 B_2 B_1) &= \rho(B_1)\rho(B_2)\rho(B_1) \\
&= R_{12} R_{23} R_{12} \\
&= R_{23} R_{12} R_{23} \quad \text{by Proposition 4.3.1} \\
&= \rho(B_2)\rho(B_1)\rho(B_2) \\
&= \rho(B_2 B_1 B_2).
\end{aligned}
$$

$\square$

Example 4.3.3. Let K be a field with char$(K) \neq 2$ and let $\langle g \rangle = C_2$ denote the cyclic group of order 2. Then (KC_2, R) is a quasitriangular two-dimensional K-Hopf algebra with quasitriangular structures $R_0 = 1 \otimes 1$ and

$$
R_1 = \frac{1}{2}(1 \otimes 1) + \frac{1}{2}(1 \otimes g) + \frac{1}{2}(g \otimes 1) - \frac{1}{2}(g \otimes g),
$$

cf. Proposition 4.1.13. Now, $(KC_2)^{\otimes^3}$ is a eight-dimensional vector space over K. Choose the basis

$$
\{1 \otimes 1 \otimes 1, 1 \otimes 1 \otimes g, 1 \otimes g \otimes 1, 1 \otimes g \otimes g, g \otimes 1 \otimes 1, g \otimes 1 \otimes g, g \otimes g \otimes 1, g \otimes g \otimes g\}
$$

for $(KC_2)^{\otimes^3}$ over K. By Proposition 4.3.2 there are two eight-dimensional linear representations of $\mathcal{B}$. The first is given by the quasitriangular Hopf algebra (KC_2, R_0) and has the form

$$
\rho_0 : \mathcal{B} \to GL_8(K),
$$

with

$$
\rho_0(B_1) = R_{12} = \begin{pmatrix}
1 & 0 & 0 & 0 & 0 & 0 & 0 & 0 \\
0 & 1 & 0 & 0 & 0 & 0 & 0 & 0 \\
0 & 0 & 0 & 0 & 1 & 0 & 0 & 0 \\
0 & 0 & 0 & 0 & 0 & 1 & 0 & 0 \\
0 & 0 & 1 & 0 & 0 & 0 & 0 & 0 \\
0 & 0 & 0 & 1 & 0 & 0 & 0 & 0 \\
0 & 0 & 0 & 0 & 0 & 0 & 1 & 0 \\
0 & 0 & 0 & 0 & 0 & 0 & 0 & 1
\end{pmatrix},
$$

and

$$
\rho_0(B_2) = R_{23} = \begin{pmatrix}
1 & 0 & 0 & 0 & 0 & 0 & 0 & 0 \\
0 & 0 & 1 & 0 & 0 & 0 & 0 & 0 \\
0 & 1 & 0 & 0 & 0 & 0 & 0 & 0 \\
0 & 0 & 0 & 1 & 0 & 0 & 0 & 0 \\
0 & 0 & 0 & 0 & 1 & 0 & 0 & 0 \\
0 & 0 & 0 & 0 & 0 & 0 & 1 & 0 \\
0 & 0 & 0 & 0 & 0 & 1 & 0 & 0 \\
0 & 0 & 0 & 0 & 0 & 0 & 0 & 1
\end{pmatrix}.
$$

The second is given by the quasitriangular Hopf algebra (KC_2, R_1) and has the form

$$\rho_1 : \mathcal{B} \to \mathrm{GL}_8(K),$$

with

$$\rho_1(B_1) = R_{12} = \begin{pmatrix} \frac{1}{2} & 0 & \frac{1}{2} & 0 & \frac{1}{2} & 0 & -\frac{1}{2} & 0 \\ 0 & \frac{1}{2} & 0 & \frac{1}{2} & 0 & \frac{1}{2} & 0 & -\frac{1}{2} \\ \frac{1}{2} & 0 & -\frac{1}{2} & 0 & \frac{1}{2} & 0 & \frac{1}{2} & 0 \\ 0 & \frac{1}{2} & 0 & -\frac{1}{2} & 0 & \frac{1}{2} & 0 & \frac{1}{2} \\ \frac{1}{2} & 0 & \frac{1}{2} & 0 & -\frac{1}{2} & 0 & \frac{1}{2} & 0 \\ 0 & \frac{1}{2} & 0 & \frac{1}{2} & 0 & -\frac{1}{2} & 0 & \frac{1}{2} \\ -\frac{1}{2} & 0 & \frac{1}{2} & 0 & \frac{1}{2} & 0 & \frac{1}{2} & 0 \\ 0 & -\frac{1}{2} & 0 & \frac{1}{2} & 0 & \frac{1}{2} & 0 & \frac{1}{2} \end{pmatrix},$$

and

$$\rho_1(B_2) = R_{23} = \begin{pmatrix} \frac{1}{2} & \frac{1}{2} & \frac{1}{2} & -\frac{1}{2} & 0 & 0 & 0 & 0 \\ \frac{1}{2} & -\frac{1}{2} & \frac{1}{2} & -\frac{1}{2} & 0 & 0 & 0 & 0 \\ \frac{1}{2} & \frac{1}{2} & -\frac{1}{2} & \frac{1}{2} & 0 & 0 & 0 & 0 \\ -\frac{1}{2} & \frac{1}{2} & \frac{1}{2} & -\frac{1}{2} & 0 & 0 & 0 & 0 \\ 0 & 0 & 0 & 0 & \frac{1}{2} & \frac{1}{2} & \frac{1}{2} & -\frac{1}{2} \\ 0 & 0 & 0 & 0 & \frac{1}{2} & -\frac{1}{2} & \frac{1}{2} & -\frac{1}{2} \\ 0 & 0 & 0 & 0 & \frac{1}{2} & \frac{1}{2} & -\frac{1}{2} & \frac{1}{2} \\ 0 & 0 & 0 & 0 & -\frac{1}{2} & \frac{1}{2} & \frac{1}{2} & -\frac{1}{2} \end{pmatrix}.$$

As one can check,

$$\rho_0(\mathcal{B}) \cong \rho_1(\mathcal{B}) \cong S_3.$$

4.4 Hopf Algebras and Affine Varieties

In this section we show how K-Hopf algebras are related to affine varieties over K. We show that an affine variety Λ with coordinate ring $K[\Lambda]$ can be identified with the collection of K-algebra maps $K[\Lambda] \to K$, and this allows us to think of the geometric object Λ as an algebraic object through the algebraic structure of its coordinate ring $K[\Lambda]$. We show that if $K[\Lambda]$ is a bialgebra, then Λ is a monoid, and if $K[\Lambda]$ is a Hopf algebra, then Λ is a group.

* * *

Let $n \geq 1$ be an integer, let K be an infinite field containing $\mathbb{Q}$, and let $K^n = \underbrace{K \times K \times \cdots \times K}_{n}$ denote the cartesian product of n copies of K. Let $x_1, x_2, \ldots, x_n$ be indeterminates and let $K[x_1, x_2, \ldots, x_n]$ denote the ring of polynomials over K.

Definition 4.4.1. An **affine variety** Λ over K is a subset Λ of K^n consisting of all common zeros $a = (a_1, a_2, \ldots, a_n)$ in K of a finite set $F = \{f_1, f_2, \ldots, f_m\}$ of polynomials in $K[x_1, x_2, \ldots, x_n]$, that is,

$$\Lambda = \{(a_1, a_2, \ldots, a_n) \in K^n : f(a) = 0, \forall f \in F\}.$$

The affine variety $\Lambda \subseteq K^n$ is the set of all simultaneous solutions of the set of equations $f_1 = 0, f_2 = 0, \ldots, f_m = 0$ if and only if $g(\Lambda) = 0$ for every g in the ideal $N = (f_1, f_2, \ldots, f_m)$ generated by $f_1, f_2, \ldots, f_m$.

Example 4.4.2. $\Lambda = \{0, 1\} \subseteq K^1$ is an affine variety since it is the set of zeros of the polynomial $f(x) = x^2 - x \in K[x]$.

Example 4.4.3. $\Lambda = \{-1, 2\} \subseteq K^1$ is an affine variety since it is the set of common zeros of the polynomials $\{x^2 - x - 2, x^3 - 4x^2 + x + 6\} \subseteq K[x]$.

Example 4.4.4. $\Lambda = \{(a, a^{-1}) : a \in K^\times\} \subseteq K^2$ is an affine variety since it is the set of zeros in K of the polynomial $f(x_1, x_2) = x_1 x_2 - 1 \in K[x_1, x_2]$.

Example 4.4.5 (Generalization of Example 4.4.4).

$$\Lambda = \{(a_{1,1}, a_{1,2}, a_{2,1}, a_{2,2}, b^{-1}) : a_{i,j} \in K, b = a_{1,1}a_{2,2} - a_{1,2}a_{2,1} \in K^\times\} \subseteq K^5$$

is an affine variety since it is the set of zeros in K of the polynomial

$$f(x_{1,1}, x_{1,2}, x_{2,1}, x_{2,2}, y) = (x_{1,1}x_{2,2} - x_{1,2}x_{2,1})y - 1$$

in $K[x_{1,1}, x_{1,2}, x_{2,1}, x_{2,2}, y]$.

Example 4.4.6. $\Lambda = K^n$ is an affine variety since it is the set of zeros of the zero polynomial $f(x_1, x_2, \ldots, x_n) = 0$.

Let $\Lambda \subseteq K^n$ be an affine variety. The set of polynomials

$$N_\Lambda = \{f \in K[x_1, x_2, \ldots, x_n] : f(a) = 0, \forall a \in \Lambda\},$$

is an ideal of $K[x_1, x_2, \ldots, x_n]$ called the **ideal of** Λ. For example, if Λ is the affine variety $\Lambda = \{(a, a^{-1}) : a \in K^\times\} \subseteq K^2$, then $N_\Lambda = (x_1 x_2 - 1)$ and if $\Lambda = K^n$, then $N_\Lambda = 0$.

Let Λ be the affine variety that consists of the common zeros of the set $\{f_1, f_2, \ldots, f_m\}$ of polynomials in $K[x_1, x_2, \ldots, x_n]$. Then $(f_1, f_2, \ldots, f_m) \subseteq N_\Lambda$. We could, of course, have proper containment. For instance, take $K = \mathbb{Q}$. Then $x^3 - 1$ defines the affine variety $\Lambda = \{1\}$ with $(x^3 - 1) \subset N_\Lambda = (x - 1)$.

By the Hilbert Basis Theorem, $K[x_1, x_2, \ldots, x_n]$ is Noetherian, see [Wat79, A.5]. Thus $N_\Lambda = (g_1, g_2, \ldots, g_l)$ for some polynomials $g_1, g_2, \ldots, g_l \in K[x_1, x_2, \ldots, x_n]$.

Proposition 4.4.7. *Let* $\Lambda \subseteq K^n$ *be an affine variety with ideal* $N_\Lambda = (g_1, g_2, \ldots, g_l)$. *Let* Λ' *denote the affine variety consisting of the common zeros in K of the polynomials* $\{g_1, g_2, \ldots, g_l\}$. *Then* $\Lambda = \Lambda'$.

Proof. By the definition of N_Λ, we have $\Lambda \subseteq \Lambda'$. Since Λ is an affine variety, it consists of points in K^n that are common zeros of some set of polynomials $\{f_1, f_2, \ldots, f_m\}$ in $K[x_1, x_2, \ldots, x_n]$. One has $(f_1, f_2, \ldots, f_m) \subseteq N_\Lambda$. Now let $b \in \Lambda'$. Then b is a zero of all polynomials in N_Λ, hence b is a common zero of the polynomials $\{f_1, f_2, \ldots, f_m\}$, and so, $b \in \Lambda$. $\qquad\square$

Let $\Lambda \subseteq K^n$ be an affine variety. A function $\phi : \Lambda \to K$ is **regular** if there exists a polynomial $f \in K[x_1, x_2, \ldots, x_n]$ for which $\phi(a) = f(a)$ for all $a \in \Lambda$. It is not hard to show that the collection of all regular functions on Λ is a ring with ring operations defined pointwise, this is the **ring of regular functions $K[\Lambda]$ on** Λ. The ring of regular functions $K[\Lambda]$ is also called the **coordinate ring** of Λ. We want to give a precise description of $K[\Lambda]$.

Proposition 4.4.8. *Let $\Lambda \subseteq K^n$ be an affine variety and let N_Λ denote the ideal of Λ. Then $K[\Lambda] = K[x_1, x_2, \ldots, x_n]/N_\Lambda$.*

Proof. Suppose that $g, h \in K[x_1, x_2, \ldots, x_n]$ with $g - h \in N_\Lambda$. Then for all $a = (a_1, a_2, \ldots, a_n) \in \Lambda$,

$$0 = (g - h)(a) = g(a) - h(a),$$

and hence $g(a) = h(a)$. Thus, g, h define the same regular functions on Λ. $\qquad\square$

For instance, if Λ is the affine variety $\Lambda = \{(a, a^{-1}) : a \in K^\times\} \subseteq K^2$, then $K[\Lambda] = K[x_1, x_2]/(x_1 x_2 - 1) = K[x, x^{-1}]$ and if $\Lambda = K^n$, then $K[\Lambda] = K[x_1, x_2, \ldots, x_n]$.

Proposition 4.4.9. *Let $\Lambda \subseteq K^n$ be an affine variety with coordinate ring $K[\Lambda]$. Then*

$$\Lambda = \mathrm{Hom}_{K\text{-}alg}(K[\Lambda], K),$$

the identification being $a = (\phi \mapsto \phi(a))$, for $a \in \Lambda$, $\phi \in K[\Lambda]$.

Proof. By Proposition 4.4.8, $K[\Lambda] = K[x_1, x_2, \ldots, x_n]/N_\Lambda$. Consequently, by [Wat79, §1.2, Theorem] $\mathrm{Hom}_{K\text{-}alg}(K[\Lambda], K)$ consists of precisely those points in K^n that are common zeros of all polynomials in N_Λ. The result then follows from Proposition 4.4.7. $\qquad\square$

It is natural to think of the affine variety $\Lambda \subseteq K^n$ in geometric terms as a collection of points in the space K^n. We can think about Λ in algebraic terms and this is done through the algebraic structure of the coordinate ring $K[\Lambda]$. We recall our identification

$$\Lambda = \mathrm{Hom}_{K\text{-}alg}(K[\Lambda], K)$$

where $a \in \Lambda$ is equated with the K-algebra homomorphism $\phi \mapsto \phi(a)$, for all $\phi \in K[\Lambda]$.

Now we assume that the coordinate ring $B = K[\Lambda]$ is not just a K-algebra but is a K-bialgebra. In that case, we obtain a monoid structure on Λ, as we now show.

Proposition 4.4.10. *There exists a binary operation Θ on $Hom_{K\text{-}alg}(B, K)$*

$$\Theta : Hom_{K\text{-}alg}(B, K) \times Hom_{K\text{-}alg}(B, K) \to Hom_{K\text{-}alg}(B, K)$$

defined as

$$\Theta(f, g)(a) = m_K(f \otimes g)\Delta_B(a) = \sum_{(a)} f(a_{(1)})g(a_{(2)}),$$

for $f, g \in Hom_{K\text{-}alg}(B, K)$, $a \in B$, $\Delta_B(a) = \sum_{(a)} a_{(1)} \otimes a_{(2)}$.

Proof. This amounts to showing that the map $\Theta(f, g) : B \to K$ is a homomorphism of K-algebras. To this end, let $a, b \in B$, $r \in K$. Then

$$
\begin{aligned}
\Theta(f, g)(ra + b) &= m_K(f \otimes g)\Delta_B(ra + b) \\
&= m_K(f \otimes g)(r\Delta_B(a) + \Delta_B(b)) \\
&= m_K(f \otimes g)\left(\left(\sum_{(a)} ra_{(1)} \otimes a_{(2)}\right) + \left(\sum_{(b)} b_{(1)} \otimes b_{(2)}\right)\right) \\
&= m_K\left(\left(\sum_{(a)} f(ra_{(1)}) \otimes g(a_{(2)})\right) + \left(\sum_{(b)} f(b_{(1)}) \otimes g(b_{(2)})\right)\right) \\
&= \left(\sum_{(a)} rf(a_{(1)})g(a_{(2)})\right) + \left(\sum_{(b)} f(b_{(1)})g(b_{(2)})\right) \\
&= r\Theta(f, g)(a) + \Theta(f, g)(b),
\end{aligned}
$$

and so $\Theta(f, g)$ is K-linear. We next show that $\Theta(f, g)$ respects multiplication, thus:

$$
\begin{aligned}
\Theta(f, g)(ab) &= m_K(f \otimes g)\Delta_B(ab) \\
&= m_K(f \otimes g)\left(\sum_{(a,b)} a_{(1)}b_{(1)} \otimes a_{(2)}b_{(2)}\right) \\
&= \sum_{(a,b)} f(a_{(1)}b_{(1)})g(a_{(2)}b_{(2)}) \\
&= \sum_{(a,b)} f(a_{(1)})f(b_{(1)})g(a_{(2)})g(b_{(2)}) \\
&= \sum_{(a,b)} f(a_{(1)})g(a_{(2)})f(b_{(1)})g(b_{(2)}) \\
&= \sum_{(a)} f(a_{(1)})g(a_{(2)}) \sum_{(b)} f(b_{(1)})g(b_{(2)}) \\
&= \Theta(f, g)(a)\Theta(f, g)(b).
\end{aligned}
$$

$\square$

Of course, as the reader may have noticed, the binary operation of Proposition 4.4.10 is precisely the convolution operation on $\mathrm{Hom}_K(B, K)$ restricted to $\mathrm{Hom}_{K\text{-alg}}(B, K) \subseteq \mathrm{Hom}_K(B, K)$, see §3.1.

Proposition 4.4.11. *Let Λ be an affine variety with coordinate ring $K[\Lambda]$. Assume that $K[\Lambda]$ is a K-bialgebra. Then $\Lambda = \mathrm{Hom}_{K\text{-alg}}(B, K)$ together with convolution $*$ is a monoid.*

Proof. By Proposition 3.1.6, $*$ is associative, and so $\mathrm{Hom}_{K\text{-alg}}(B, K)$ is a semigroup. The composition $\lambda_K \epsilon_B : B \to K$ serves as a two-sided identity element with respect to $*$, thus:

$$(f * \lambda_K \epsilon_B)(a) = \sum_{(a)} f(a_{(1)}) \lambda_K(\epsilon_B(a_{(2)}))$$

$$= \sum_{(a)} \lambda_K(\epsilon_B(a_{(2)})) f(a_{(1)})$$

$$= \sum_{(a)} \epsilon_B(a_{(2)}) f(a_{(1)})$$

$$= \sum_{(a)} f(\epsilon_B(a_{(2)}) a_{(1)})$$

$$= f\left(\sum_{(a)} \epsilon_B(a_{(2)}) a_{(1)} \right)$$

$$= f(a) \quad \text{by the counit property.}$$

In a similar manner one obtains $\lambda_K \epsilon_B * f = f$. Thus $\mathrm{Hom}_{K\text{-alg}}(B, K)$ is a monoid under $*$. $\square$

We next assume that the coordinate ring $K[\Lambda]$ is a K-Hopf algebra H.

Proposition 4.4.12. *$\mathrm{Hom}_{K\text{-alg}}(H, K)$ together with convolution $*$ is a group.*

Proof. By Proposition 4.4.11 $\mathrm{Hom}_{K\text{-alg}}(H, K)$ together with $*$ is a monoid with two-sided identity $\lambda_K \epsilon_H$. So we only need to show that each element $f \in \mathrm{Hom}_{K\text{-alg}}(H, K)$ has a two-sided inverse under $*$. Note that $f\sigma_H \in \mathrm{Hom}_{K\text{-alg}}(H, K)$. Now, for all $a \in H$,

$$(f\sigma_H * f)(a) = \sum_{(a)} f(\sigma_H(a_{(1)})) f(a_{(2)})$$

$$= f\left(\sum_{(a)} \sigma_H(a_{(1)}) a_{(2)} \right),$$

since f is an algebra homomorphism. Continuing with the calculation:

$$f\left(\sum_{(a)} \sigma_H(a_{(1)})a_{(2)}\right) = f(\epsilon_H(a)1_H) \quad \text{by the coinverse property}$$

$$= \epsilon_H(a)1_K$$

$$= \lambda_K(\epsilon_H(a))$$

$$= (\lambda_K\epsilon_H)(a).$$

In a similar manner one obtains $f * f\sigma_H = \lambda_K\epsilon_H$.   □

The conclusion of Proposition 4.4.12 is the following: If a given K-Hopf algebra H is the coordinate ring of an affine variety $\Lambda \subseteq K^n$ (necessarily, H is a commutative K-algebra), then $\langle \Lambda, * \rangle$ with $\Lambda = \text{Hom}_{K\text{-alg}}(H, K)$ is a group. In this way we can put a group structure on the geometric object Λ. Here are some examples.

Example 4.4.13. The polynomial ring $K[x]$ is a K-Hopf algebra with x primitive. As one can check, it is the coordinate ring of the affine variety K^1. Thus $\langle K^1, * \rangle$ is a group, where in this case, $a * b = a + b$ for all $a, b \in K^1$; $K^1 = \text{Hom}_{K\text{-alg}}(K[x], K)$ is the **additive group** of K.

Example 4.4.14. $K[x, x^{-1}]$ with x group-like is a K-Hopf algebra and it is the coordinate ring of the affine variety $\Lambda = \{(a, a^{-1}) : a \in K^{\times}\}$. Now, $a * b = ab$, and so, $\langle \Lambda, * \rangle$ is the **multiplicative group** of non-zero elements of K.

Example 4.4.15. $K[x_{1,1}, x_{1,2}, x_{2,1}, x_{2,2}, 1/(x_{1,1}x_{2,2} - x_{1,2}x_{2,1})]$ is a K-Hopf algebra (§4.6, Exercise 15). It is the coordinate ring of the affine variety Λ of Example 4.4.5. An element $a \in \Lambda$ can be viewed as the invertible matrix

$$M = \begin{pmatrix} a_{1,1} & a_{1,2} \\ a_{2,1} & a_{2,2} \end{pmatrix}.$$

Convolution $*$ is now ordinary matrix multiplication and $\langle \Lambda, * \rangle$ is the group of matrices $GL_2(K)$.

Example 4.4.16. $\mathbb{C}[x]/(x^3 - 1)$ is an $\mathbb{C}$-Hopf algebra with $\bar{x}$ group-like; it is the coordinate ring of the affine variety $\Lambda = \{1, \zeta_3, \zeta_3^2\} \subseteq \mathbb{C}^1$; $\langle \Lambda, * \rangle \cong C_3$, in this case.

Note: If H is a commutative K-Hopf algebra and K is replaced with any commutative K-algebra A, then $\text{Hom}_{K\text{-alg}}(H, A)$ is a group, see [Un11, Proposition 3.1.5]. More generally, if H is a K-Hopf algebra (commutative or non-commutative) and A is a commutative K-algebra, then $\text{Hom}_{K\text{-alg}}(H, A)$ is a group [Ab77, Theorem 2.1.5].

4.5 Hopf Algebras and Hopf Galois Extensions

In this section we show that L is a Galois extension of K with group G if and only if L is a Galois KG-extension of K. In this latter form (L is a Galois KG-extension) the concept of a classical Galois extension L/K can be extended to rings of integers. Ultimately, Galois KG-extensions (and hence, classical Galois extensions) can be generalized to Galois H-extensions of rings S/R where H is a Hopf algebra.

$$* \quad * \quad *$$

Let K be a finite extension of $\mathbb{Q}$ and let L be a finite extension of K. Let $\text{Aut}_K(L)$ denote the group of automorphisms of L that fix K and let G be a subgroup of $\text{Aut}_K(L)$. Let S denote the ring of integers of L. Now, L is a KG-module with scalar multiplication given as

$$\left(\sum_{g \in G} a_g g \right) \cdot x = \sum_{g \in G} a_g g(x),$$

for $a_g \in K, x \in L$.

Now, of course, KG is a K-Hopf algebra.

Proposition 4.5.1. *Let L be a finite extension of K and let $G \leq \text{Aut}_K(L)$. Then L is a KG-module algebra.*

Proof. For each $g \in G$ one has

$$g \cdot (xy) = g(xy) = g(x)g(y) = (g \cdot x)(g \cdot y),$$

and

$$g \cdot 1_L = g(1_L) = 1_L = \epsilon_{KG}(g)1_L,$$

for all $x, y \in L$. Thus for $h = \sum_{g \in G} a_g g$,

$$h \cdot (xy) = \sum_{g \in G} a_g g(xy)$$

$$= \sum_{g \in G} a_g (g \cdot x)(g \cdot y)$$

$$= \sum_{(h)} (h_{(1)} \cdot x)(h_{(2)} \cdot y),$$

and

$$h \cdot 1_L = \sum_{g \in G} a_g g(1_L) = \sum_{g \in G} a_g 1_L = \epsilon_{KG}(h)1_L.$$

$\square$

The field L is a vector space over K and we consider

$$\mathrm{End}_K(L) = \mathrm{Hom}_K(L, L)$$

the collection of K-linear transformations $\phi : L \to L$. Over K, $\mathrm{End}_K(L)$ is a vector space with addition given pointwise:

$$(\phi + \psi)(x) = \phi(x) + \psi(x),$$

and scalar multiplication defined as $(r\phi)(x) = r\phi(x)$ for $r \in K$, $x \in L$. As a set of automorphisms of L that fix K, the group G is a subset of $\mathrm{End}_K(L)$. There is a homomorphism of vector spaces

$$J : L \otimes_K KG \to \mathrm{End}_K(L) \tag{4.18}$$

defined as $J(x \otimes h)(y) = x(h \cdot y)$ for $x, y \in L$, $h \in KG$.

Lemma 4.5.2. *The elements of G form a linearly independent set of vectors over L.*

Proof. Write the elements of G as $g_0, g_1, g_2, \ldots, g_{n-1}$. If the set $\{g_0, g_1, \ldots, g_{n-1}\}$ is not linearly independent over L, then there exists a smallest positive integer m, $1 \le m \le n$, and a set of distinct integers $i_1, i_2, \ldots, i_m$, $0 \le i_1, i_2, \ldots, i_m \le n - 1$ for which

$$a_1 g_{i_1} + a_2 g_{i_2} + \cdots + a_{m-1} g_{i_{m-1}} + a_m g_{i_m} = 0 \tag{4.19}$$

with $a_1, a_2, \ldots, a_m$ non-zero. Since $g_{i_{m-1}} \ne g_{i_m}$, there exists $y \in L$ for which $g_{i_{m-1}}(y) \ne g_{i_m}(y)$ with $g_{i_m}(y) \ne 0$. For any $x \in L$,

$$a_1 g_{i_1}(xy) + \cdots + a_{m-1} g_{i_{m-1}}(xy) + a_m g_{i_m}(xy) = 0$$

thus

$$a_1 g_{i_1}(y) g_{i_1}(x) + \cdots + a_{m-1} g_{i_{m-1}}(y) g_{i_{m-1}}(x) + a_m g_{i_m}(y) g_{i_m}(x) = 0,$$

and so,

$$a_1 g_{i_1}(y) g_{i_1} + \cdots + a_{m-1} g_{i_{m-1}}(y) g_{i_{m-1}} + a_m g_{i_m}(y) g_{i_m} = 0. \tag{4.20}$$

Now dividing (4.20) by $g_{i_m}(y)$ yields

$$a_1 \left(\frac{g_{i_1}(y)}{g_{i_m}(y)} \right) g_{i_1} + \cdots + a_{m-1} \left(\frac{g_{i_{m-1}}(y)}{g_{i_m}(y)} \right) g_{i_{m-1}} + a_m g_{i_m} = 0,$$

and subtracting (4.19) gives

$$\left(a_1\left(\frac{g_{i_1}(y)}{g_{i_m}(y)}\right) - a_1\right)g_{i_1} + \cdots + \left(a_{m-1}\left(\frac{g_{i_{m-1}}(y)}{g_{i_m}(y)}\right) - a_{m-1}\right)g_{i_{m-1}} = 0.$$

Note that $a_{m-1}\left(\frac{g_{i_{m-1}}(y)}{g_{i_m}(y)}\right) - a_{m-1} \neq 0$ and so, we have a contradiction of the minimality of m. Thus G is linearly independent. $\qquad\square$

We give a characterization of Galois extensions.

Proposition 4.5.3. *Let K be a finite extension of $\mathbb{Q}$, let L be a finite extension of K, and let G be a subgroup of $\mathrm{Aut}_K(L)$. Then the map*

$$J : L \otimes_K KG \to \mathrm{End}_K(L)$$

is a bijection if and only if L is a Galois extension of K with group G.

Proof. Suppose L is a Galois extension of K with group G. We show that J is a bijection. Let $h = \sum_{i=0}^{n-1} a_i g_i \in KG$, $x \in L$, and suppose that

$$J\left(x \otimes \sum_{i=0}^{n-1} a_i g_i\right)(y) = \left(\sum_{i=0}^{n-1} x a_i g_i\right) \cdot y = \sum_{i=0}^{n-1} x a_i g_i(y) = 0,$$

for all $y \in L$. By Lemma 4.5.2 $\{g_0, g_1, \ldots, g_{n-1}\}$ is linearly independent over L. Thus $x a_i = 0$ for all i, and so J is an injection.

Since L/K is Galois, $|G| = [L : K]$ and so

$$\dim(L \otimes_K KG) = [L : K]^2 = \dim(\mathrm{End}_K(L)).$$

Thus J is surjective.

For the converse, suppose that J is a bijection. Then $[L : K]|G| = [L : K]^2$, and so, $|G| = [L : K]$. Put $L = K(\alpha)$ with $p(x) = \mathrm{irr}(\alpha, K)$ of degree $[L : K]$. Since each element of G moves α to some distinct root β of $p(x)$, L is the splitting field of $p(x)$ over K, thus L/K is Galois. Let $\mathrm{Gal}(L/K)$ denote the Galois group of L over K. We have $G \leq \mathrm{Gal}(L/K)$. But since $|\mathrm{Gal}(L/K)| = [L : K] = |G|$, $G = \mathrm{Gal}(L/K)$ and L is Galois over K with group G. $\qquad\square$

We have shown the following: the notion that L is a Galois extension of K with group G is equivalent to L being a KG-module algebra for which the map

$$J : L \otimes_K KG \to \mathrm{End}_K(L),$$

defined as $J(x \otimes h)(y) = x(h \cdot y)$ for $x, y \in L$, $h \in KG$ is a bijection. We say the L is a "Galois KG-extension of K," or: KG is "realizable as a Galois group." Of course, KG is not a group—it is a Hopf algebra—but this terminology makes sense since KG does correspond to the group $\mathrm{Hom}_{K\text{-alg}}(KG, K)$ as in §4.4.

In this latter form (involving the bijection $\jmath$) the notion of a classical Galois extension L/K can be generalized to rings of integers. Let S be the ring of integers of L, let R be the ring of integers of K; S will play the role of L; RG will play the role of KG. Certainly, S is an RG-module since $x \in S$ implies that $g(x) \in S$ for all $g \in G$. In fact, S is an RG-module algebra (proof?)

Definition 4.5.4. Let L be a finite Galois extension of K with group G. Let S be the ring of integers of L, let R be the ring of integers of K. Then S is a **Galois RG-extension of R** if the map

$$\jmath : S \otimes_R RG \to \mathrm{End}_R(S),$$

defined as $\jmath(x \otimes h)(y) = x(h \cdot y)$ for $x, y \in S$, $h \in RG$ is an isomorphism of R-modules. We also say that RG is **realizable as a Galois group**.

The challenge is to construct a Galois RG-extension. Here is a criterion that we can use.

Proposition 4.5.5. *Let L be a Galois extension of K with group G. Then S is a Galois RG-extension of R if and only if every prime ideal P of R is unramified in S.*

Proof. The proof is beyond the scope of this book. The interested reader should see [Ch00, §2, p. 18].   □

The ring of integers R of a finite extension K of $\mathbb{Q}$ with $[K : \mathbb{Q}] > 1$ can *never* be a Galois $\mathbb{Z}G$-extension since at least one prime p in $\mathbb{Z}$ is ramified in R [Ne99, Theorem III.2.17].

So in order to construct Galois RG-extensions with $|G| > 1$, we must start with a base field K that is larger than $\mathbb{Q}$. Our Galois RG-extension will be the ring of integers of the **Hilbert Class Field** of K, which is the maximal abelian unramified extension of K. The Hilbert Class Field of K is a Galois extension with Galois group isomorphic to the class group of K, cf. [IR90, Notes, p. 184].

Proposition 4.5.6. *Let K be a finite extension of $\mathbb{Q}$ for which R is not a PID, that is, for which the class number $h_R > 1$. Then there exists a Galois extension L of K with group G for which the ring of integers S is a Galois RG-extension of R.*

Proof. Choose L to be the Hilbert Class Field of K. Then each prime P of R is unramified in S, and so by Proposition 4.5.5, S is a Galois RG-extension of R.   □

Example 4.5.7. Let $K = \mathbb{Q}(\sqrt{-5})$ with $R = \mathbb{Z}[\sqrt{-5}]$. The class number of R is 2. The Hilbert Class Field of K is $L = K(i)$, and the ring of integers of L is $S = R[(i + \sqrt{-5})/2]$. By Proposition 4.5.5, S is a Galois RC_2-extension of R.

For more examples of unramified (hence Galois) extensions of number fields, see [He66].

We notice that in the Galois H-extensions given above, with $H = KG$ or $H = RG$, H is a Hopf algebra and the H-module algebra structure of L is given by the classical Galois action of G on L. We can broaden the notion of Galois extension to include module algebras over Hopf algebras in which the action of the Hopf algebra is not given by the classical Galois action.

Definition 4.5.8. Let R be a commutative ring with unity, let H be a cocommutative R-Hopf algebra which is finitely generated and projective as an R-module, and let S be a commutative R-algebra which is finitely generated and projective as an R-module. Then S is a **Galois H-extension of R** if S is an H-module algebra with action denoted as $h \cdot y$ for $h \in H$, $y \in S$, and the map

$$J : S \otimes_R H \to \text{End}_R(S),$$

defined as $J(x \otimes h)(y) = x(h \cdot y)$ for $x, y \in S$, $h \in H$ is an isomorphism of R-modules. We also say that H is **realizable as a Galois group** and that S is a **Hopf Galois extension** of R.

Here is perhaps the easiest example of a Hopf Galois extension that will yield a non-classical Galois action, cf. [CS69, pp. 35–39].

Example 4.5.9 (S. Chase, M. Sweedler). Let C_n be the cyclic group of order n, generated by g. Let R be a commutative ring with unity and let a be a unit of R. Then the R-algebra $R[z]$ with $z^n = a$, is a C_n-graded R-algebra, that is,

$$R[z] = R \oplus Rz \oplus Rz^2 \oplus \cdots \oplus Rz^{n-1},$$

with $Rz^i Rz^j \subseteq Rz^{i+j}$ for $0 \leq i, j \leq n - 1$. Let RC_n be the group ring R-Hopf algebra with linear dual RC_n^* and let $\{p_i\}_{i=0}^{n-1}$ denote the dual basis for RC_n^*, $p_i(g^j) = \delta_{i,j}$. As one can check, $R[z]$ is a left RC_n^*-module algebra with action $p_i(z^j) = \delta_{i,j} z^j$. The map

$$J : R[z] \otimes_R RC_n^* \to \text{End}_R(R[z]),$$

defined as $J(z^i \otimes p_j)(z^k) = z^i p_j(z^k) = \delta_{i,j} z^{i+k}$ is a bijection, and so, $R[z]$ is a Galois RC_n^*-extension of R.

More generally, for G a finite group, a G-graded R-algebra $A = \oplus_{\sigma \in G} A_\sigma$ is a left RG^*-module algebra, and A is a Galois RG^*-extension if and only if $A_e = R$ and A is strongly graded, that is, $A_\sigma \cdot A_\tau = A_{\sigma\tau}$ for all $\sigma, \tau \in G$, see [Ca98, Proposition 8.2.1]. The terminology "strongly graded" is due to Dade [Da80].

We have the following special case of Example 4.5.9.

Example 4.5.10. Let $n \geq 2$, let K be any field containing $\mathbb{Q}$, and let a be an element of K that is not an nth power of an element of K. Then $p(x) = x^n - a$ is irreducible over K. Let α be a zero of $p(x)$ in $\mathbb{C}$. Then $L = K(\alpha)$ is a simple algebraic extension of K of degree n, and L is a C_n-graded K-algebra and a Galois KC_n^*-extension of K with Galois action $p_i(\alpha^j) = \delta_{i,j}\alpha^j$.

Example 4.5.11. In the case $K = \mathbb{Q}$, $n > 2$, of Example 4.5.10, we have that $\mathbb{Q}(\alpha)$ is a $\mathbb{Q}C_n^*$-Galois extension of $\mathbb{Q}$; the $\mathbb{Q}C_n^*$-module algebra structure of $\mathbb{Q}(\alpha)$ is not the classical Galois action; $\mathbb{Q}(\alpha)$ is not a Galois extension of $\mathbb{Q}$.

A finite extension of fields $K/\mathbb{Q}$ can have more than one Hopf Galois structure.

Example 4.5.12. Let $p(x) = x^3 - 2$ and let K be the splitting field of $p(x)$ over $\mathbb{Q}$. Then $K = \mathbb{Q}(\zeta_3, \sqrt[3]{2})$ and K is a Galois extension of $\mathbb{Q}$ with group $S_3 = \langle \sigma, \tau \rangle$ defined by the relations $\sigma^3 = \tau^2 = 1$, $\tau\sigma = \sigma^2\tau$. Now, by Proposition 4.5.3, K is a Galois $\mathbb{Q}S_3$-extension of $\mathbb{Q}$.

We want to put a non-classical Hopf Galois structure on K. Consider the group ring KS_3. Let S_3 act on KS_3 as the Galois group on K and by conjugation on S_3; denote this action by "·." Thus, for $a_i \in K$,

$$\sigma \cdot (a_0 + a_1\sigma + a_2\sigma^2 + a_3\tau + a_4\sigma\tau + a_5\sigma^2\tau)$$

$$= \sigma(a_0) + \sigma(a_1)\sigma + \sigma(a_2)\sigma^2 + \sigma(a_3)\sigma^2\tau + \sigma(a_4)\tau + \sigma(a_5)\sigma\tau$$

and

$$\tau \cdot (a_0 + a_1\sigma + a_2\sigma^2 + a_3\tau + a_4\sigma\tau + a_5\sigma^2\tau)$$

$$= \tau(a_0) + \tau(a_1)\sigma^2 + \tau(a_2)\sigma + \tau(a_3)\tau + \tau(a_4)\sigma^2\tau + \tau(a_5)\sigma\tau.$$

Now let

$$KS_3^{S_3} = \{x \in KS_3 : \sigma(x) = \tau(x) = x\}$$

denote the subset of KS_3 fixed by S_3. Then,

$$KS_3^{S_3} = \{b_0 + b_1\sigma + \tau(b_1)\sigma^2 + b_2\tau + \sigma^2(b_2)\sigma\tau + \sigma(b_2)\sigma^2\tau\},$$

where $b_0 \in \mathbb{Q}$, $b_1 \in \mathbb{Q}(\zeta_3)$ and $b_2 \in \mathbb{Q}(\sqrt[3]{2})$.

$H = KS_3^{S_3}$ is a six-dimensional $\mathbb{Q}$-Hopf algebra which acts on K as a module algebra through ·, and K is a Galois H-extension of $\mathbb{Q}$. One has $H \not\cong \mathbb{Q}S_3$ and so, the Hopf Galois structure on K is different than the classical Galois structure.

For more examples of separable, non-Galois extensions of $\mathbb{Q}$ which have (or don't have) Hopf Galois structures on them, see [CRV14a, CRV14b] and [CRV14c].

In Example 4.5.10 we take $n = p$, a prime number, $K = \mathbb{Q}(\zeta_{p^m})$, for a fixed integer $m \geq 1$, and $a = \zeta_{p^m}$. Then $p(x) = x^p - \zeta_{p^m}$ is irreducible over K with zero $\alpha = \zeta_{p^{m+1}}$. $L = K(\zeta_{p^{m+1}})$ is a KC_p^*-Galois extension of K. By Proposition 3.1.14, $KC_p^* \cong KC_p$ and so L is a KC_p-Galois extension of K with the KC_p-module algebra action on L induced from the classical Galois action of C_p on L. (Indeed, L is a Galois extension K with group C_p.)

For $K = \mathbb{Q}(\zeta_{p^m})$ we have $R = \mathbb{Z}[\zeta_{p^m}]$ and the ideal (p) has unique factorization

$$(p) = (1 - \zeta_{p^m})^{p^{m-1}(p-1)}.$$

In this case, $e' = p^{m-1}$. Let $\lambda = \zeta_{p^m} - 1$. We know from §3.4 that there is a collection of Hopf orders in KC_p of the form

$$H(i) = R\left[\frac{g-1}{\lambda^i}\right], \ \langle g \rangle = C_p,$$

for $0 \le i \le e'$. We ask: which (if any) of these R-Hopf orders are realizable as Galois groups, that is, for which $H(i)$ does there exist a degree p extension L/K whose ring of integers S is an $H(i)$-Galois extension of R?

Let i be an integer $1 \le i \le e'$, let $i' = e' - i$, and let

$$w_{i'} = 1 + \lambda^{pi'+1}.$$

Then $p(x) = x^p - w_{i'}$ is irreducible over K. The splitting field of $p(x)$ over K is $L_{i'} = K(\sqrt[p]{w_{i'}})$; $L_{i'}$ is a Galois extension of K with group C_p, generated by

$$g : \sqrt[p]{w_{i'}} \mapsto \zeta_p \sqrt[p]{w_{i'}}.$$

Let $S_{i'}$ be the ring of integers of $L_{i'}$.

Take $i = e'$, so that $i' = 0$. Then

$$w_0 = 1 + (\zeta_{p^m} - 1) = \zeta_{p^m}.$$

One has

$$L_0 = K(\zeta_{p^{m+1}}) = \mathbb{Q}(\zeta_{m+1}).$$

with $S_0 = R[\zeta_{p^{m+1}}] = \mathbb{Z}[\zeta_{p^{m+1}}]$.

Proposition 4.5.13. *Assume the notation as above. Then S_0 is a Galois $H(e')$-extension of R. That is, the R-Hopf order $H(e')$ in KC_p is realizable as a Galois group.*

Proof. We apply Example 4.5.9 of Chase and Sweedler. S_0 is a C_p-graded R-algebra with $\zeta_{p^{m+1}}^p = \zeta_{p^m}$, a unit in R. Thus S_0 is an RC_p^*-Galois extension of R. But note that $RC_p^* \cong H(e')$ by Proposition 3.4.11. $\square$

What about the remaining R-Hopf orders $H(i)$ for $0 \le i < e'$? It is not so clear which of these (if any) are realizable as Galois groups. They are all realizable as Galois groups if we localize.

Again, suppose that $K = \mathbb{Q}(\zeta_{p^n})$, $L_{i'} = K(\sqrt[p]{w_{i'}})$, with

$$w_{i'} = 1 + (\zeta_{p^n} - 1)^{pi'+1},$$

$0 \le i \le e'$, $e' = p^{n-1}$. Let R be the ring of integers of K, let $S_{i'}$ be the ring of integers of $L_{i'}$. Let

$$(p) = Q_1^{e_1} Q_2^{e_2} \cdots Q_m^{e_m}$$

be the unique factorization of (p) in $S_{i'}$. Let $Q = Q_1$ and let $(L_{i'})_Q$ denote the completion of $L_{i'}$ at Q with valuation ring $(S_{i'})_Q$; let $P = (\lambda)$ and let K_P be the completion of K at P with valuation ring R_P and uniformizing parameter π. There is a collection of R_P-Hopf orders in $K_P C_p$ of the form

$$H_P(i) = R_P \left[\frac{g-1}{\pi^i} \right],$$

for $0 \le i \le e'$. Now, Childs [Ch87] has shown that each $H_P(i)$ is realizable as a Galois group; Childs also computes the ring of integers that realizes $H_P(i)$.

Proposition 4.5.14. *Assume the notation as above. Then for each integer* i, $0 \le i \le e'$,

(i) $(S_{i'})_Q = \hat{R}_P[x]$ *with* $x = (\sqrt[q]{w_{i'}} - 1)/\pi^{i'}$,
(ii) $(S_{i'})_Q$ *is a Galois* $H_P(i)$*-extension of* R_P, *in other words,* $(S_{i'})_Q$ *is an* $H_P(i)$*- module algebra and the map*

$$j : (S_{i'})_Q \otimes_{R_P} H_P(i) \to End_{R_P}((S_{i'})_Q),$$

defined as $j(x \otimes h)(y) = xh(y)$ *for* $x, y \in (S_{i'})_Q$, $h \in H_P(i)$, *is a* R_P*-module isomorphism.*

Proof. See [Ch87, §14, Proposition 14.3].   □

4.6 Chapter Exercises

Exercises for §4.1

1. Let (B, R) be an almost commutative K-bialgebra with $R = \sum_{i=1}^{n} a_i \otimes b_i$. Let I be a biideal of B. Prove that $(B/I, \overline{R})$ with $\overline{R} = \sum_{i=1}^{n} (a_i + I) \otimes (b_i + I)$ is almost commutative.

2. In the proof of Proposition 4.1.10 take $z = 0$ and compute the image of R' under the inverse map $\phi^{-1} : KT^* \to KT$. Does $R = \phi^{-1}(R')$ satisfy the quasitriangular conditions (4.2) or (4.3) with $B = KT$?

3. Let (B, R) be a quasitriangular K-bialgebra with $R = \sum_{i=1}^{n} a_i \otimes b_i$. Let $f : B^* \to B$ be the map defined as $f(\alpha) = (\alpha \otimes I_B)(R) = \sum_{i=1}^{n} \alpha(a_i)b_i$ for $\alpha \in B^*$.

   (a) Prove that f is a homomorphism of K-algebras.
   (b) Assuming that B is finite dimensional, prove that f is a coalgebra anti-homomorphism.

4. Let (H, R) be a quasitriangular K-Hopf algebra with $R = \sum_{i=1}^{n} a_i \otimes b_i$. Let I be a Hopf ideal of H. Prove that $(H/I, \overline{R})$ with

$$\overline{R} = \sum_{i=1}^{n} (a_i + I) \otimes (b_i + I)$$

is quasitriangular.

5. Let H denote M. Sweedler's Hopf algebra of Example 3.1.5. Let

$$R = \frac{1}{2}(1 \otimes 1) + \frac{1}{2}(1 \otimes g) + \frac{1}{2}(g \otimes 1) - \frac{1}{2}(g \otimes g)$$
$$+ x \otimes x - x \otimes gx + gx \otimes x + gx \otimes gx.$$

Prove that (H, R) is a quasitriangular Hopf algebra.

Exercises for §4.2

6. Draw the braid associated with the braid product $B_2 B_1^2 B_2^{-1}$.
7. Draw the braids that verify the braid relation $B_1 B_2 B_1 = B_2 B_1 B_2$.
8. Let $S = \{B_1^2, B_2^2\}$ be a subset of the braid group $\mathcal{B}$ and let $\langle S \rangle$ denote the subgroup of $\mathcal{B}$ generated by S.

   (a) Prove that $\langle S \rangle \triangleleft \mathcal{B}$.
   (b) Compute $\mathcal{B}/\langle S \rangle$.

9. Decompose the braid

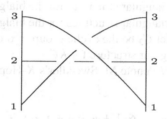

   into a product of fundamental braids.

Exercises for §4.3

10. Compute the regular representation $\rho : \mathcal{V} \to GL_4(K)$ of the Klein 4-group $\mathcal{V}$.
11. Let $\rho_1 : \mathcal{B} \to GL_8(K)$ denote the representation of the braid group given in Example 4.3.3. Prove that $\rho_1(\mathcal{B}) \cong S_3$.

Exercises for §4.4

12. Let B be a K-bialgebra, let A be a K-algebra, and let $\text{Hom}_K(B, A)$ denote the collection of K-linear maps $B \to A$. Show that $\langle \text{Hom}_K(B, A), * \rangle$ is a monoid.
13. Let K be a field and let $A = K[x]/(x^n)$, $n \geq 2$. Show that A is not the coordinate ring of an affine variety $X \subseteq K^1$. What if $n = 1$?
14. Let $A = \mathbb{Q}[x_1, x_2]/(2 - x_1^3 + x_2^2)$. Prove that A is the coordinate ring of an affine variety $X \subseteq \mathbb{Q}^2$.
15. Let K be a field and let $K[x_{1,1}, x_{1,2}, x_{2,1}, x_{2,2}, y]$ be the K-algebra of polynomials in the variables $x_{1,1}, x_{1,2}, x_{2,1}, x_{2,2}, y$. Let

$$f(x_{1,1}, x_{1,2}, x_{2,1}, x_{2,2}, y) = (x_{1,1} x_{2,2} - x_{1,2} x_{2,1})y - 1.$$

   Show that

$$K[x_{1,1}, x_{1,2}, x_{2,1}, x_{2,2}, y]/(f(x_{1,1}, x_{1,2}, x_{2,1}, x_{2,2}, y))$$

is a K-Hopf algebra. *Hint: Define comultiplication as*

$$\Delta(x_{i,j}) = \sum_{k=1}^{2} x_{i,k} \otimes x_{k,j}$$

for $1 \le i, j \le 2$.

Exercises for §4.5

16. Let $K = \mathbb{Q}(\zeta_{p^n})$, $e' = p^{n-1}$.
    (a) Find all integers i, $0 \le i \le e'$, for which $w = 1 + (\zeta_{p^n} - 1)^{pi+1}$ is a unit in $R = \mathbb{Z}[\zeta_{p^n}]$.
    (b) For any i with $0 \le i \le e'$ satisfying (a), prove that the ring of integers S of $L = K(\sqrt[p]{w})$ is a Galois $H(i')$-extension of R, $i = e' - i$.

Questions for Further Study

1. Find all of the quasitriangular structures for the bialgebra $K[x]$ of Example 2.1.3.
2. Find all of the quasitriangular structures for the bialgebra $K[x]$ of Example 2.1.4.
3. Let $K = \mathbb{Q}(\zeta_3)$ and let C_3 be the cyclic group of order 3 generated by g. Find a non-trivial quasitriangular structure for KC_3.
4. Let $K = Z_3$ and let H denote M. Sweedler's K-Hopf algebra of Example 3.1.5. Let

$$R = -1 \otimes 1 - 1 \otimes g - g \otimes 1 + g \otimes g + x \otimes x - x \otimes gx + gx \otimes x + gx \otimes gx.$$

Then (H, R) is quasitriangular (§4.6, Exercise 5). Compute $\rho(B_1)$ where $\rho : \mathcal{B} \to \mathrm{GF}_{64}(Z_3)$ is the representation given by R.
5. Referring to Example 4.5.12, let $K = \mathbb{Q}(\zeta_3, \sqrt[3]{2})$ and let

$$H = KS_3^{S_3} = \{b_0 + b_1\sigma + \tau(b_1)\sigma^2 + b_2\tau + \sigma^2(b_2)\sigma\tau + \sigma(b_2)\sigma^2\tau\},$$

where $b_0 \in \mathbb{Q}$, $b_1 \in \mathbb{Q}(\zeta_3)$ and $b_2 \in \mathbb{Q}(\sqrt[3]{2})$.

    (a) Prove that H is a six-dimensional $\mathbb{Q}$-Hopf algebra.
    (b) Show that K is a Galois H-extension of $\mathbb{Q}$.
    (c) Obtain the Wedderburn–Malcev decomposition of both H and $\mathbb{Q}S_3$. Conclude that $H \not\cong \mathbb{Q}S_3$.

Bibliography

[Ab77] E. Abe, *Hopf Algebras* (Cambridge University of Press, Cambridge, 1977)

[Bi93] J.S. Birman, New points of view in knot theory. Bull. Am. Math. Soc. **28**(2), 253–287 (1993)

[Br82] K. Brown, *Cohomology of Groups* (Springer, New York, 1982)

[By93a] N. Byott, Cleft extensions of Hopf algebras I. J. Algebra **157**, 405–429 (1993)

[By93b] N. Byott, Cleft extensions of Hopf algebras, II. Proc. Lond. Math. Soc. **67**, 227–307 (1993)

[Ca98] S. Caenepeel, *Brauer Groups, Hopf Algebras and Galois Theory*. K-Monographs in Mathematics (Kluwer, Dordrecht, 1998)

[CF67] J.W.S. Cassels, A. Frohlich (eds.), *Algebraic Number Theory* (Academic, London, 1967)

[CS69] S.U. Chase, M. Sweedler, *Hopf Algebras and Galois Theory*. Lecture Notes in Mathematics, vol. 97 (Springer, Berlin, 1969)

[Ch79] L.N. Childs, *A Concrete Introduction to Higher Algebra* (Springer, New York, 1979)

[Ch87] L.N. Childs, Taming wild extensions with Hopf algebras. Trans. Am. Math. Soc. **304**, 111–140 (1987)

[Ch00] L.N. Childs, *Taming Wild Extensions: Hopf Algebras and Local Galois Module Theory*. Mathematical Surveys and Monographs, vol. 80 (American Mathematical Society, Providence, RI, 2000)

[CU03] L.N. Childs, R.G. Underwood, Cyclic Hopf orders defined by isogenies of formal groups. Am. J. Math. **125**, 1295–1334 (2003)

[CU04] L.N. Childs, R.G. Underwood, Duals of formal group Hopf orders in cyclic groups. Ill. J. Math. **48**(3), 923–940 (2004)

[CG93] W. Chin, J. Goldman, Bialgebras of linearly recursive sequences. Commun. Algebra **21**(11), 3935–3952 (1993)

[CRV14a] T. Crespo, A. Rio, M. Vela, From Galois to Hopf Galois: theory and practice. Contemp. Math. (to appear). arXiv:1403.6300, (2014)

[CRV14b] T. Crespo, A. Rio, M. Vela, On the Galois correspondence theorem in separable Hopf Galois theory. arXiv:1405.0881, (2014)

[CRV14c] T. Crespo, A. Rio, M. Vela, The Hopf Galois property in subfield lattices. arXiv:1309.5754, (2014)

[Da80] E. Dade, Group-graded rings and modules. Math. Z. **174**, 241–262 (1980)

[Dr86] V.G. Drinfeld, Quantum groups, in *Proceedings of International Congress of Mathematics*, Berkeley, CA, vol. 1, 1986, pp. 789–820

© Springer International Publishing Switzerland 2015
R.G. Underwood, *Fundamentals of Hopf Algebras*, Universitext,
DOI 10.1007/978-3-319-18991-8

[Dr90] V.G. Drinfeld, On almost commutative Hopf algebras. Leningrad Math. J. **1**, 321–342 (1990)

[Ei74] S. Eilenberg, *Automata, Languages, and Machines*, vol. A (Academic, New York, 1974)

[Eis95] D. Eisenbud, *Commutative Algebra* (Springer, New York, 1995)

[FT91] A. Frohlich, M.J. Taylor, *Algebraic Number Theory* (Cambridge University Press, Cambridge, 1991)

[Go97] F.Q. Gouvêa, *p-Adic Numbers* (Springer, Berlin, 1997)

[Gr92] C. Greither, Extensions of finite group schemes, and Hopf Galois theory over a complete discrete valuation ring. Math. Z. **210**, 37–67 (1992)

[He66] C.S. Herz, Construction of class fields, in *Seminar on Complex Multiplication*. Lecture Notes in Mathematics, vol. 21, VII-1-VII-21, (Springer, Berlin, 1966)

[Hi40] G. Higman, The units of group-rings. Proc. Lond. Math. Soc. (2) **46**, 231–248 (1940)

[HS71] P.J. Hilton, U. Stammbach, *A Course in Homological Algebra* (Springer-Verlag, New York, 1971)

[HK71] K. Hoffman, R. Kunze, *Linear Algebra*, 2nd edn. (Prentice-Hall, Englewood Cliffs, NJ, 1971)

[Ho64] F.E. Hohn, *Elementary Matrix Algebra*, 2nd edn. (Macmillan, New York, 1964)

[HU79] J.E. Hopcroft, J.D. Ulman, *Introduction to Automata Theory, Languages, and Computation* (Addison-Wesley, Reading, MA, 1979)

[IR90] K. Ireland, M. Rosen, *A Classical Introduction to Modern Number Theory*, 2nd edn. (Springer, New York, 1990)

[Ja51] N. Jacobson, *Lectures in Abstract Algebra* (D. Van Nostrand, Princeton, NJ, 1951)

[La84] S. Lang, *Algebra*, 2nd edn. (Addison-Wesley, Reading , MA, 1984)

[La86] S. Lang, *Algebraic Number Theory* (Springer, New York, 1986)

[Lar67] R.G. Larson, Group rings over Dedekind domains. J. Algebra **5**, 358–361 (1967)

[LS69] R.G. Larson, M.E. Sweedler, An associative orthogonal bilinear form for Hopf algebras. Am. J. Math. **91**, 75–93 (1969)

[Lar88] R.G. Larson, Hopf algebras via symbolic algebra, in *Computers in Algebra*. Lecture Notes in Pure and Applied Mathematics, vol. 111 (Dekker, New York, 1988), pp. 91–97

[LT90] R. Larson, E. Taft, The algebraic structure of linearly recursive sequences under Hadamard product. Israel J. Math. **72**, 118–132 (1990)

[LN97] R. Lidl, H. Niederreiter, *Finite Fields* (Cambridge University Press, Cambridge, 1997)

[Ma04] W. Mao, *Modern Cryptography* (Prentice Hall PTR, Upper Saddle River, NJ, 2004)

[Mar82] G.E. Martin, *Transformation Geometry* (Springer, New York, 1982)

[Mi96] J.S. Milne, Class field theory, unpublished class notes (1996)

[Mo93] S. Montgomery, *Hopf Algebras and Their Actions on Rings*. CBMS Regional Conference Series in Mathematics, vol. 82 (American Mathematical Society, Providence, RI, 1993)

[Mu75] J. Munkres, *Topology* (Prentice-Hall, NJ, 1975)

[Ne99] J. Neukirch, *Algebraic Number Theory*. Grundlehren der mathematischen Wissenschaften, vol. 322 (Springer, Berlin, 1999)

[Ni12] F. Nichita, Introduction to the Yang-Baxter equation with open problems. Axioms **1**, 33–37 (2012)

[NU11] W. Nichols, R. Underwood, Algebraic Myhill-Nerode theorems. Theor. Comput. Sci. **412**, 448–457 (2011)

[Ra93] D.E. Radford, Minimal quasitriangular Hopf algebras. J. Algebra **157**, 285–315 (1993)

[Ro02] J. Rotman, *Advanced Modern Algebra* (Pearson, Upper Saddle River, NJ, 2002)

[Ru76] W. Rudin, *Principles of Mathematical Analysis*, 3rd edn. (McGraw-Hill, New York, 1976)

[Sa08] P. Samuel, *Algebraic Theory of Numbers* (Dover, New York, 2008)

[Se13] S.K. Sehgal, Units of integral group rings-a survey. Retrieved online: www.math.ualberta.ca/people/Faculty/Sehgal/publications/057.pdf. 7 Aug 2013

[Sh74] I.R. Shafarevich, *Basic Algebraic Geometry* (Springer, Berlin, 1974)

[Si85] P. Singh, The so-called fibonacci numbers in ancient and medieval India. Hist. Math. **12**(3), 229–244 (1985)

[Sw69] M.E. Sweedler, *Hopf Algebras* (W. A. Benjamin, New York, 1969)

[Ta94] E. Taft, Algebraic aspects of linearly recursive sequences, in *Advances in Hopf Algebras*, ed. by J. Bergen, S. Montgomery (Marcel Dekker, New York, 1994)

[TO70] J. Tate, F. Oort, Group schemes of prime order. Ann. Sci. Ec. Norm. Sup. **3**, 1–21 (1970)

[TB92] M.J. Taylor, N.P. Byott, Hopf orders and Galois module structure, in *DMV Seminar*, vol. 18 (Birkhauser Verlag, Basel, 1992), pp. 154–210

[Un94] R. Underwood, R-Hopf algebra orders in KC_{p^2}. J. Algebra **169**, 418–440 (1994)

[Un96] R. Underwood, The valuative condition and R algebra orders in KC_{p^3}. Am. J. Math. **118**, 401–743 (1996)

[Un98] R. Underwood, The structure and realizability of R-Hopf orders in KC_{p^3}. Commun. Algebra **26**(11), 3447–3462 (1998)

[Un99] R. Underwood, Isogenies of polynomial formal groups. J. Algebra **212**, 428–459 (1999)

[UC05] R. Underwood, L.N. Childs, Duality for Hopf orders. Trans. Am. Math. Soc. **358**(3), 1117–1163 (2006)

[Un06] R. Underwood, Realizable Hopf orders in KC_8. Int. Math. Forum **1**(17–20), 833–851 (2006)

[Un08a] R. Underwood, Sequences of Hopf-Swan subgroups. J. Number Theory **128**(7), 1900–1915 (2008)

[Un08b] R. Underwood, Realizable Hopf orders in KC_{p^2}. J. Algebra **319**(11), 4426–4455 (2008)

[Un11] R. Underwood, *An Introduction to Hopf Algebras* (Springer, New York/Dordrecht/Heidelberg/London, 2011)

[Un12] R. Underwood, Quasitriangular structure of Myhill-Nerode bialgebras. Axioms **1**(2), 155–172 (2012). doi:10.3390/axioms1020155

[Wa97] L.C. Washington, *Introduction to Cyclotomic Fields* (Springer, New York, 1997)

[Wat79] W. Waterhouse, *Introduction to Affine Group Schemes* (Springer, New York, 1979)

[We63] E. Weiss, *Algebraic Number Theory* (Chelsea, New York, 1963)

[Wen90] H. Wenzl, Representations of braid groups and the quantum Yang-Baxter equation. Pac. J. Math. **145**(1), 153–180 (1990)

[ZM73] N. Zierler, W.H. Mills, Products of linear recurring sequences. J. Algebra **27**, 147–157 (1973)

Index

© Springer International Publishing Switzerland 2015
R.G. Underwood, *Fundamentals of Hopf Algebras*, Universitext,
DOI 10.1007/978-3-319-18991-8

Printed in the United States
By Bookmasters